SpringerBriefs in Statistics

JSS Research Series in Statistics

The current research of statistics in Japan has expanded in several directions in line with recent trends in academic activities in the area of statistics and statistical sciences over the globe. The core of these research activities in statistics in Japan has been the Japan Statistical Society (JSS). This society, the oldest and largest academic organization for statistics in Japan, was founded in 1931 by a handful of pioneer statisticians and economists and now has a history of about 80 years. Many distinguished scholars have been members, including the influential statistician Hirotugu Akaike, who was a past president of JSS, and the notable mathematician Kiyosi Itô, who was an earlier member of the Institute of Statistical Mathematics (ISM), which has been a closely related organization since the establishment of ISM. The society has two academic journals: the Journal of the Japan Statistical Society (English Series) and the Journal of the Japan Statistical Society (Japanese Series). The membership of JSS consists of researchers, teachers, and professional statisticians in many different fields including mathematics, statistics, engineering, medical sciences, government statistics, economics, business, psychology, education, and many other natural, biological, and social sciences. The JSS Series of Statistics aims to publish recent results of current research activities in the areas of statistics and statistical sciences in Japan that otherwise would not be available in English; they are complementary to the two JSS academic journals, both English and Japanese. Because the scope of a research paper in academic journals inevitably has become narrowly focused and condensed in recent years, this series is intended to fill the gap between academic research activities and the form of a single academic paper. The series will be of great interest to a wide audience of researchers, teachers, professional statisticians, and graduate students in many countries who are interested in statistics and statistical sciences, in statistical theory, and in various areas of statistical applications.

Shonosuke Sugasawa · Tatsuya Kubokawa

Mixed-Effects Models and Small Area Estimation

Springer

Shonosuke Sugasawa
Center for Spatial Information Science
University of Tokyo
Kashiwa-shi, Chiba, Japan

Tatsuya Kubokawa
Faculty of Economics
University of Tokyo
Tokyo, Japan

ISSN 2191-544X ISSN 2191-5458 (electronic)
SpringerBriefs in Statistics
ISSN 2364-0057 ISSN 2364-0065 (electronic)
JSS Research Series in Statistics
ISBN 978-981-19-9485-2 ISBN 978-981-19-9486-9 (eBook)
https://doi.org/10.1007/978-981-19-9486-9

This Springer imprint is published by the registered company Springer Nature Singapore Pte Ltd.
The registered company address is: 152 Beach Road, #21-01/04 Gateway East, Singapore 189721, Singapore

Preface

This book provides a self-contained introduction of mixed-effects models and small area estimation techniques. In particular, it focuses on both introducing classical theory and reviewing the latest methods. It first introduces basic issues of mixed-effects models, such as parameter estimation, random effects prediction, variable selection, and asymptotic theory. Standard mixed-effects models used in small area estimation, known as Fay–Herriot model and nested error regression model, are then introduced. Both frequentist and Bayesian approaches are given to compute predictors of small area parameters of interest. For measuring uncertainty of the predictors, several methods to calculate mean squared errors and confidence intervals are discussed. Various advanced approaches using mixed-effects models are introduced, covering from frequentist to Bayesian approaches. This book is helpful for researchers and graduate students in various fields requiring data analysis skills as well as in mathematical statistics.

The authors would like to thank Professor Masafumi Akahira for giving us the opportunity of publishing this book. The work of the first author was supported in part by Grant-in-Aid for Scientific Research (21H00699) from the Japan Society for the Promotion of Science (JSPI). The work of the second author was supported in part by Grant-in-Aid for Scientific Research (18K11188) from the JSPI.

Tokyo, Japan
September 2022

Shonosuke Sugasawa
Tatsuya Kubokawa

Contents

Chapter 1
Introduction

The term 'small area' or 'small domain' refers to a small geographical region such as a county, municipality or state, or a small demographic group such as a specific age–sex–race group. In the estimation of a characteristic of such a small group, the direct estimate based on only on the data from the small group is likely to be unreliable, because only the small number of observations are available from the small group. The problem of small area estimation is how to produce a reliable estimate for the characteristic of the small group, and the small area estimation has been actively and extensively studied from both theoretical and practical aspects due to an increasing demand for reliable small area estimates from public and private sectors. The articles by Ghosh and Rao (1994) and Pfeffermann (2013) give good reviews and motivations, and the comprehensive book by Rao and Molina (2015) covers all the main developments in small area estimation. More recent review on the use of mixed models in small area estimation is given in Sugasawa and Kubokawa (2020). Also see Demidenko (2004) for general mixed models and Pratesi (2016) for analysis of poverty data by small area estimation. In this paper, we describe the details of classical methods and give a review of recent developments, which will be helpful for readers who are interested in this topic.

To improve the accuracy of direct survey estimates, we make use of the relevant supplementary information such as data from other related areas and covariate data from other sources. The linear mixed models (LMM) enable us to 'borrow strength' from the relevant supplementary data, and the resulting model-based estimators or the best linear unbiased predictors (BLUP) provide reliable estimates for the small area characteristics. The BLUP shrinks the direct estimates in small areas toward a stable quantity constructed by pooling all the data, thereby BLUP is characterized by the effects of pooling and shrinkage of the data. These two features of BLUP mainly come from the structure of linear mixed models described as (observation) = (common parameters) + (random effects) + (error terms), namely, the shrinkage

S. Sugasawa and T. Kubokawa, *Mixed-Effects Models and Small Area Estimation*, JSS Research Series in Statistics, https://doi.org/10.1007/978-981-19-9486-9_1

effect arises from the random effects, and the pooling effect is due to the setup of the common parameters. While BLUP was originally proposed by Henderson (1950), empirical version of BLUP (EBLUP) is related to the classical shrinkage estimator studied by Stein (1956), who established analytically that EBLUP improves on the sample means when the number of small areas is larger than or equal to three. This fact shows not only that EBLUP has a larger precision than the sample mean, but also that a similar concept came out at the same time by Henderson (1950) for practical use and Stein (1956) for theoretical interest. Based on these historical backgrounds, there have been a lot of methods proposed so far.

As a former part of this book, we first introduce details of theory of linear mixed models and BLUP (or EBLUP) in Chap. 2. Since measuring the variability or risk of EBLUP is an important task in small area estimation, we focus on mean squared errors and prediction intervals in Chap. 3, and describe several methods based on asymptotic calculations and simulation-based methods such as jackknife and bootstrap. Our argument tries to keep generality without assuming normality assumptions for the distribution as much as possible. We then introduce two basic small area models, the Fay–Herriot model (Fay and Herriot 1979) and the nested error regression (Battese et al. 1988) in Chap. 4. In Chap. 5, we provide basic techniques of hypothesis testing and variable selection in linear mixed models, which can immediately be applied to the basic small area models.

As a latter part of this book, we focus more on techniques of small area estimation based on mixed-effects models, with some examples in places. First, in Chap. 6, we explain advanced theory of basic small area models to handle practical problems. We mainly focus on three techniques, adjusted likelihood methods for estimating random effects variance, observed best prediction for random effects and robust prediction and fitting of the small area models. In Chap. 7, we introduce several techniques to handle non-normal response variables in small area estimation, including generalized linear mixed models, models with data transformation, and models with non-normal distributions. Finally, we review several extensions of the basic small area models in Chap. 8. The topics treated there are flexible modeling of random effects, measurement error models, nonparametric and semiparametric models, and heteroscedastic variance models.

References

Battese G, Harter R, Fuller W (1988) An error-components model for prediction of county crop areas using survey and satellite data. J Am Stat Assoc 83:28–36

Demidenko E (2004) Mixed models: theory and applications. Wiley

Fay R, Herriot R (1979) Estimators of income for small area places: An application of James-Stein procedures to census. J Am Stati Assoc 74:341–353

Ghosh M, Rao J (1994) Small area estimation: an appraisal. Stat Sci 9:55–76

Henderson C (1950) Estimation of genetic parameters. Ann Math Stat 21:309–310

Pfeffermann D (2013) New important developments in small area estimation. Stat Sci 28:40–68

Pratesi M (ed) (2016) Analysis of poverty data by small area estimation. Wiley

Rao JNK, Molina I (2015) Small area estimation, 2nd edn. Wiley
Stein C (1956) Inadmissibility of the usual estimator for the mean of a multivariate normal distribution. Proc Third Berkeley Symp Math Stat Probab 1:197–206
Sugasawa S, Kubokawa T (2020) Small area estimation with mixed models: a review. Japanese J Stat Data Sci 3:693–720

Chapter 2
General Mixed-Effects Models and BLUP

Linear mixed models are widely used in a variety of scientific areas such as small area estimation (Rao and Molina 2015), longitudinal data analysis (Verbeke and Molenberghs 2006), and meta-analysis (Boreinstein et al. 2009), and estimation of variance components plays an essential role in fitting the models. In this chapter, we provide the general mixed-effects models, some examples, and the derivation of the best linear unbiased predictors. For estimating unknown parameters like variance components, we suggest the general estimating equations which include the restricted maximum likelihood estimators and derive their asymptotic properties.

2.1 Mixed-Effects Models and Examples

Consider the general linear mixed model

$$\boldsymbol{y} = \boldsymbol{X}\boldsymbol{\beta} + \boldsymbol{Z}\boldsymbol{v} + \boldsymbol{\epsilon}, \tag{2.1}$$

where $\boldsymbol{y}$ is an $N \times 1$ observation vector of the response variable; $\boldsymbol{X}$ and $\boldsymbol{Z}$ are $N \times p$ and $N \times m$ matrices, respectively, of the explanatory variables; $\boldsymbol{\beta}$ is a $p \times 1$ unknown vector of the regression coefficients; $\boldsymbol{v}$ is an $m \times 1$ vector of the random effects; and $\boldsymbol{\epsilon}$ is an $N \times 1$ vector of the random errors. Here, $\boldsymbol{v}$ and $\boldsymbol{\epsilon}$ are mutually independently distributed as $\mathrm{E}[\boldsymbol{v}] = \boldsymbol{0}$, $\mathrm{E}[\boldsymbol{v}\boldsymbol{v}^\top] = \boldsymbol{R}_v(\boldsymbol{\psi})$, $\mathrm{E}[\boldsymbol{\epsilon}] = \boldsymbol{0}$, and $\mathrm{E}[\boldsymbol{\epsilon}\boldsymbol{\epsilon}^\top] = \boldsymbol{R}_e(\boldsymbol{\psi})$, where $\boldsymbol{\psi} = (\psi_1, \ldots, \psi_q)^\top$ is a q-dimensional vector of unknown parameters, and $\boldsymbol{R}_v = \boldsymbol{R}_v(\boldsymbol{\psi})$ and $\boldsymbol{R}_e = \boldsymbol{R}_e(\boldsymbol{\psi})$ are positive definite matrices. Throughout the paper, for simplicity, it is assumed that $\boldsymbol{X}$ is of full rank. Then, the mean and the covariance matrices of $\boldsymbol{y}$ are $\mathrm{E}[\boldsymbol{y}] = \boldsymbol{X}\boldsymbol{\beta}$ and

$$\mathbf{Cov}\,(\boldsymbol{y}) = \boldsymbol{\Sigma} = \boldsymbol{\Sigma}(\boldsymbol{\psi}) = \boldsymbol{R}_e(\boldsymbol{\psi}) + \boldsymbol{Z}\boldsymbol{R}_v(\boldsymbol{\psi})\boldsymbol{Z}^\top. \tag{2.2}$$

S. Sugasawa and T. Kubokawa, *Mixed-Effects Models and Small Area Estimation*, JSS Research Series in Statistics, https://doi.org/10.1007/978-981-19-9486-9_2

When $\boldsymbol{v}$ and $\boldsymbol{\epsilon}$ have multivariate normal distributions $\boldsymbol{v} \sim \mathrm{N}(\mathbf{0}, \boldsymbol{R}_v)$ and $\boldsymbol{\epsilon} \sim \mathrm{N}(\mathbf{0}, \boldsymbol{R}_e)$, the model (2.1) is expressed as the Bayesian model

$$\begin{aligned} \boldsymbol{y} \mid \boldsymbol{\mu} &\sim \mathrm{N}(\boldsymbol{\mu}, \boldsymbol{R}_e), \\ \boldsymbol{\mu} &\sim \mathrm{N}(\boldsymbol{X}\boldsymbol{\beta}, \boldsymbol{Z}\boldsymbol{R}_v\boldsymbol{Z}^\top). \end{aligned} \tag{2.3}$$

The general linear mixed model includes several specific models used in applications. Some of them are given below.

Example 2.1 (Fay–Herriot model) The Fay–Herriot (FM) model is a basic area-level model which is useful in the small area estimation. Let y_i be a statistic for estimating a characteristic of the i-th area like the sample mean. Assume that y_i has the simple linear mixed model

$$y_i = \boldsymbol{x}_i^\top \boldsymbol{\beta} + v_i + \varepsilon_i, \quad i = 1, \dots, m, \tag{2.4}$$

where m is the number of small areas; $\boldsymbol{x}_i$ is a $p \times 1$ vector of explanatory variables; $\boldsymbol{\beta}$ is a $p \times 1$ unknown common vector of regression coefficients; and v_i's and ε_i's are mutually independent random errors distributed as $\mathrm{E}[v_i] = \mathrm{E}[\varepsilon_i] = 0$, $\mathrm{Var}(v_i) = \psi$, and $\mathrm{Var}(\varepsilon_i) = D_i$. Let $\boldsymbol{X} = (\boldsymbol{x}_1, \dots, \boldsymbol{x}_m)^\top$, $\boldsymbol{y} = (y_1, \dots, y_m)^\top$, and let $\boldsymbol{v}$ and $\boldsymbol{\epsilon}$ be similarly defined. Then, the model is expressed as

$$\boldsymbol{y} = \boldsymbol{X}\boldsymbol{\beta} + \boldsymbol{v} + \boldsymbol{\epsilon},$$

where $\mathrm{E}[\boldsymbol{y}] = \boldsymbol{X}\boldsymbol{\beta}$ and $\mathbf{Cov}(\boldsymbol{y}) = \boldsymbol{\Sigma} = \psi \boldsymbol{I}_m + \boldsymbol{D}$ for $\boldsymbol{D} = \mathrm{diag}(D_1, \dots, D_m)$ and N corresponds to m. The prediction of $\theta_i = \boldsymbol{x}_i^\top \boldsymbol{\beta} + v_i$ is of interest. □

Example 2.2 (Nested error regression model) The nested error regression (NER) model or random intercept model is a basic unit-level model used in the small area estimation. This model is described as

$$y_{ij} = \boldsymbol{x}_{ij}^\top \boldsymbol{\beta} + v_i + \varepsilon_{ij}, \quad i = 1, \dots, m, \ j = 1, \dots, n_i, \tag{2.5}$$

where m is the number of small areas; $N = \sum_{i=1}^m n_i$, $\boldsymbol{x}_{ij}$ is a $p \times 1$ vector of explanatory variables; $\boldsymbol{\beta}$ is a $p \times 1$ unknown common vector of regression coefficients; and v_i's and ε_{ij}'s are mutually independently distributed as $\mathrm{E}[v_i] = \mathrm{E}[\varepsilon_{ij}] = 0$, $\mathrm{Var}(v_i) = \tau^2$, and $\mathrm{Var}(\varepsilon_{ij}) = \sigma^2$. Here, τ^2 and σ^2 are referred to as, respectively, 'between' and 'within' components of variance, and both are unknown. Let $\boldsymbol{X}_i = (\boldsymbol{x}_{i1}, \dots, \boldsymbol{x}_{i,n_i})^\top$, $\boldsymbol{X} = (\boldsymbol{X}_1^\top, \dots, \boldsymbol{X}_m^\top)^\top$, $\boldsymbol{y}_i = (y_{i1}, \dots, y_{i,n_i})^\top$, $\boldsymbol{y} = (\boldsymbol{y}_1^\top, \dots, \boldsymbol{y}_m^\top)^\top$ and let $\boldsymbol{\epsilon}_i$ and $\boldsymbol{\epsilon}$ be similarly defined. Let $\boldsymbol{v} = (v_1, \dots, v_m)^\top$ and $\boldsymbol{Z} = \text{block diag}(\boldsymbol{j}_{n_1}, \dots, \boldsymbol{j}_{n_m})$ for $\boldsymbol{j}_k = (1, \dots, 1)^\top \in \mathbb{R}^k$. Then, the model is expressed in vector notations as $\boldsymbol{y}_i = \boldsymbol{X}_i\boldsymbol{\beta} + v_i\boldsymbol{j}_{n_i} + \boldsymbol{\epsilon}_i$ for $i = 1, \dots, m$, or $\boldsymbol{y} = \boldsymbol{X}\boldsymbol{\beta} + \boldsymbol{Z}\boldsymbol{v} + \boldsymbol{\epsilon}$.

Battese et al. (1988) used the NER model in the framework of a finite population model to predict areas under corn and soybeans for each of $m = 12$ counties in North-

Central Iowa. In their analysis, each county is divided into about 250 ha segments, and n_i segments are selected from the i-th county. For the j-th segment of the i-th county, y_{ij} is the number of hectares of corn (or soybeans) in the (i, j) segment reported by interviewing farm operators, and x_{ij1} and x_{ij2} are the number of pixels (0.45 ha) classified as corn and soybeans, respectively, by using LANDSAT satellite data. Since n_i's range from 1 to 5 with $\sum_{i=1}^{m} n_i = 37$, the sample mean $\overline{y}_i = \sum_{j=1}^{n_i} y_{ij}/n_i$ has large deviation for predicting the mean crop hectare per segment $\theta_i = \overline{\boldsymbol{x}}_j^\top \boldsymbol{\beta} + v_i$ for $\overline{\boldsymbol{x}}_i = \sum_{j=1}^{n_i} \boldsymbol{x}_{ij}/n_i$. The NER model enables us to construct more reliable prediction procedures not only by using the auxiliary information on the LANDSAT data, but also by combining the data of the related areas. □

Example 2.3 (Variance components model) The nested error regression model is extended to the variance components (VC) model suggested by Henderson (1950), which is described as

$$\boldsymbol{y} = \boldsymbol{X}\boldsymbol{\beta} + \boldsymbol{Z}_1\boldsymbol{v}_1 + \cdots + \boldsymbol{Z}_k\boldsymbol{v}_k + \boldsymbol{\epsilon}, \tag{2.6}$$

where $\boldsymbol{Z}_i$ is an $N \times r_i$ matrix, $\boldsymbol{v}_i$ is an $r_i \times 1$ random vector with $\mathrm{E}[\boldsymbol{v}_i] = \boldsymbol{0}$ and $\mathbf{Cov}\,(\boldsymbol{v}_i) = \tau_i^2 \boldsymbol{I}_{r_i}$, and $\boldsymbol{\eta}$ is an $N \times 1$ random vector with $\mathrm{E}[\boldsymbol{\epsilon}] = \boldsymbol{0}$ and $\mathbf{Cov}\,(\boldsymbol{\epsilon}) = \sigma^2 \boldsymbol{V}_0$ for known matrix $\boldsymbol{V}_0$. All the random vectors are mutually independent. The variance parameters $\tau_1^2, \ldots, \tau_k^2, \sigma^2$ are called the variance components.

The VC model includes various random effects models. For example, the two-way classification model, given by $y_{ijk} = \mu + v_{1i} + v_{2j} + \varepsilon_{ijk}$, and the two-way crossed classification model, given by $y_{ijk} = \mu + v_{1i} + v_{2j} + v_{3ij} + \varepsilon_{ijk}$, for $i = 1, \ldots, m$, $j = 1, \ldots, \ell$ and $k = 1, \ldots, n_{ij}$ belong to the VC model, where μ is an unknown mean and $\mathrm{Var}(v_{1i}) = \tau_1^2$, $\mathrm{Var}(v_{2j}) = \tau_2^2$, $\mathrm{Var}(v_{3ij}) = \tau_3^2$, and $\mathrm{Var}(\varepsilon_{ijk}) = \sigma^2$.

When the i-th area is subdivided into ℓ_i subareas, the two-fold subarea-level model is

$$y_{ij} = \boldsymbol{x}_{ij}^\top \boldsymbol{\beta} + v_i + u_{ij} + \varepsilon_{ij}, \quad j = 1, \ldots, \ell_i, \quad i = 1, \ldots, m,$$

where v_i, u_{ij}, and ε_{ij} are mutually independent random variables with $\mathrm{E}[v_i] = \mathrm{E}[u_{ij}] = \mathrm{E}[\varepsilon_{ij}] = 0$, $\mathrm{Var}(v_i) = \tau_1^2$, $\mathrm{Var}(u_{ij}) = \tau_2^2$, and $\mathrm{Var}(\varepsilon_{ij}) = D_{ij}$. The two-fold nested error regression model is also described for unit-level observations as

$$y_{ijk} = \boldsymbol{x}_{ijk}^\top \boldsymbol{\beta} + v_i + u_{ij} + \varepsilon_{ijk}, \quad k = 1, \ldots, n_{ij}, \quad j = 1, \ldots, \ell_i, \quad i = 1, \ldots, m,$$

where the distributions of v_i and u_{ij} are the same as in the two-fold subarea-level model and ε_{ij} has a distribution with $\mathrm{Var}(\varepsilon_{ij}) = \sigma^2$. □

Example 2.4 (Random coefficients model) The random coefficient (RC) model incorporates random effects in regression coefficients. This model is described as

$$y_{ij} = \boldsymbol{x}_{ij}^\top (\boldsymbol{\beta} + \boldsymbol{v}_i) + \varepsilon_{ij}, \quad i = 1, \ldots, m, \ j = 1, \ldots, n_i, \tag{2.7}$$

where $\boldsymbol{v}_1, \ldots, \boldsymbol{v}_m$ are mutually independently distributed as $\mathrm{E}[\boldsymbol{v}_i] = \boldsymbol{0}$ and $\mathbf{Cov}\,(\boldsymbol{v}_i) = \boldsymbol{\Psi}(\tau_1^2, \ldots, \tau_s^2)$, and $\mathrm{E}[\varepsilon_{ij}] = 0$ and $\mathrm{Var}(\varepsilon_{ij}) = \sigma^2$. In the nested error regression (NER) model, the regression coefficients are fixed and the intercept term is only the random effect. The RC model uses random slopes depending on areas. As a structure of the covariance matrix of $\boldsymbol{v}_i$, a simple setup is $\boldsymbol{\Psi} = \tau^2 \boldsymbol{I}_p$, but it may be realistic to take a form like $\boldsymbol{\Psi} = \mathrm{diag}\,(\tau_1^2, \ldots, \tau_p^2)$ so that variances vary depending on explanatory variables. □

Example 2.5 (Spatial model) The FH model assumes that $v_1, \ldots, v_m$ are mutually independently and identically distributed. When there exist spatial correlations, however, we need to assume a structure for correlation among v_i's. Two typical setups of the correlation structures are the conditional autoregression (CAR) spatial model and the simultaneously autoregressive (SAR) model. We here assume the normality for v_i's.

Let A_i be a set of neighboring areas of area i. The CAR model assumes that the conditional distribution of v_i given all v_ℓ for $\ell \neq i$ is

$$v_i \mid \{v_\ell : \ell \neq i\} \sim \mathrm{N}\Big(\rho \sum_{\ell \in A_i} q_{i\ell} v_\ell, \tau^2\Big),$$

which implies that $\boldsymbol{v} \sim \mathrm{N}_m(\boldsymbol{0}, \tau^2(\boldsymbol{I}_m - \rho \boldsymbol{Q})^{-1})$, where $(\boldsymbol{Q})_{i\ell} = q_{i\ell}$ and $\boldsymbol{Q}$ is a symmetric matrix with $q_{ii} = 0$ and $q_{i\ell} = 0$ for $\ell \notin A_i$. The SAR model assumes that $\boldsymbol{v} = \lambda \boldsymbol{W}\boldsymbol{v} + \boldsymbol{u}$ for $\boldsymbol{u} \sim \mathrm{N}(\boldsymbol{0}, \tau^2 \boldsymbol{I}_m)$, which is equivalently rewritten as

$$\boldsymbol{v} = (\boldsymbol{I}_m - \lambda \boldsymbol{W})^{-1}\boldsymbol{u} \sim \mathrm{N}\Big(\boldsymbol{0}, \tau^2\big\{(\boldsymbol{I}_m - \lambda \boldsymbol{W})(\boldsymbol{I}_m - \lambda \boldsymbol{W})^\top\big\}^{-1}\Big),$$

where $\boldsymbol{I}_m - \lambda \boldsymbol{W}$ is assumed to be nonsingular. The matrices $\boldsymbol{Q}$ and $\boldsymbol{W}$ describe the neighborhood structures of areas. □

Example 2.6 (Time-series cross-sectional model) The area-level model with time-series or longitudinal structures is described by

$$y_{it} = \boldsymbol{x}_{it}^\top \boldsymbol{\beta} + v_{it} + \varepsilon_{it}, \quad i = 1, \ldots, m, \quad t = 1, \ldots, T, \tag{2.8}$$

where m is the number of small areas, t is a time index, $N = mT$, $\boldsymbol{x}_{it}$ is a $p \times 1$ vector of explanatory variables, $\boldsymbol{\beta}$ is a $p \times 1$ unknown common vector of regression coefficients, and v_{it}'s and ε_{it}'s are random errors. Let $\boldsymbol{X}_i = (\boldsymbol{x}_{i1}, \ldots, \boldsymbol{x}_{i,T})^\top$, $\boldsymbol{y}_i = (y_{i1}, \ldots, y_{i,T})^\top$, and let $\boldsymbol{v}_i$ and $\boldsymbol{\epsilon}_i$ be similarly defined. Then, the model is expressed in vector notations as

$$\boldsymbol{y}_i = \boldsymbol{X}_i \boldsymbol{\beta} + \boldsymbol{v}_i + \boldsymbol{\epsilon}_i, \quad i = 1, \ldots, m.$$

Here, it is assumed that $\boldsymbol{\epsilon}_i$ and $\boldsymbol{v}_i$ are mutually distributed as $\mathrm{E}[\boldsymbol{\epsilon}_i] = \mathrm{E}[\boldsymbol{v}_i] = \boldsymbol{0}$, $\mathbf{Cov}\,(\boldsymbol{\epsilon}_i) = \boldsymbol{D}_i$ for $\boldsymbol{D}_i = \mathrm{diag}\,(d_{i1}, \ldots, d_{iT})$, and $\mathbf{Cov}\,(\boldsymbol{v}_i) = \psi \boldsymbol{\Psi}(\rho)$ for unknown scalar ψ and a positive definite matrix $\boldsymbol{\Psi}(\rho)$ with a unknown parameter ρ, $|\rho| < 1$. Two typical cases are the longitudinal structure and the autoregressive AR(1) structure, which correspond, respectively, to

$$\boldsymbol{\Psi}(\rho) = (1-\rho)\boldsymbol{I}_T + \rho \boldsymbol{j}_T \boldsymbol{j}_T^\top \quad \text{and} \quad \boldsymbol{\Psi}(\rho) = \frac{1}{1-\rho^2}\mathbf{mat}_{i,j}(\rho^{|i-j|}),$$

where the notation $\mathbf{mat}_{i,j}(\cdot)$ is defined in (2.9). Letting $\boldsymbol{X} = (\boldsymbol{X}_1^\top, \dots, \boldsymbol{X}_m^\top)^\top$, $\boldsymbol{y} = (\boldsymbol{y}_1^\top, \dots, \boldsymbol{y}_m^\top)^\top$ and letting $\boldsymbol{v}$ and $\boldsymbol{\epsilon}$ be defined similarly, we can express the model as $\boldsymbol{y} = \boldsymbol{X}\boldsymbol{\beta} + \boldsymbol{v} + \boldsymbol{\epsilon}$.

Rao and Yu (1994) suggested the different time-series cross-sectional model

$$\begin{aligned} y_{it} &= \boldsymbol{x}_{it}^\top \boldsymbol{\beta} + v_i + u_{it} + \varepsilon_{it}, \quad i = 1, \dots, m, \quad t = 1, \dots, T, \\ u_{it} &= \rho u_{i,t-1} + e_{it}, \quad |\rho| < 1, \end{aligned}$$

where ε_{it} is a sampling error with $\mathrm{E}[\varepsilon_{it}] = 0$ and $\mathrm{Var}(\varepsilon_{it}) = d_{it}$, and u_{it} and e_{it} are random errors with $\mathrm{E}[v_i] = \mathrm{E}[e_{it}] = 0,$, $\mathrm{Var}(v_i) = \tau^2$, and $\mathrm{Var}(e_{it}) = \sigma^2$. □

We finally describe the notations used in this monograph. The (a, b)-th element of matrix $\boldsymbol{V}$ and the inverse $\boldsymbol{V}^{-1}$ are denoted by $(\boldsymbol{V})_{ab}$ and $(\boldsymbol{V})^{ab}$. The partial differential operator ∂_a is defined by $\partial_a = \partial/\partial\psi_a$ for $\boldsymbol{\psi} = (\psi_1, \dots, \psi_q)^\top$. When $\boldsymbol{V}$ is a matrix of functions of $\boldsymbol{\psi}$, we use the simple notations $\boldsymbol{V}_{(a)} = \partial_a \boldsymbol{V}$ and $\boldsymbol{V}_{(ab)} = \partial_a \partial_b \boldsymbol{V}$ for $a, b = 1, \dots, q$. We also use the notations of the column vector $\mathbf{col}_i(a_i)$ and the matrix $\mathbf{mat}_{ij}(b_{ij})$ defined by

$$\mathbf{col}_i(a_i) = \begin{pmatrix} a_1 \\ \vdots \\ a_q \end{pmatrix}, \quad \mathbf{mat}_{ij}(b_{ij}) = \begin{pmatrix} b_{11} & \cdots & b_{1q} \\ \vdots & \ddots & \vdots \\ b_{q1} & \cdots & b_{qq} \end{pmatrix}. \tag{2.9}$$

2.2 Best Linear Unbiased Predictors

An interesting quantity we want to estimate is a linear combination of $\boldsymbol{\beta}$ and $\boldsymbol{v}$. Let

$$\theta = \boldsymbol{c}_1^\top \boldsymbol{\beta} + \boldsymbol{c}_2^\top \boldsymbol{v}, \tag{2.10}$$

for vectors of constants $\boldsymbol{c}_1 \in \mathbb{R}^p$ and $\boldsymbol{c}_2 \in \mathbb{R}^M$. In the example of the Fay–Herriot model, the interesting quantities are the conditional means $\mathrm{E}[y_a \mid v_a] = \boldsymbol{x}_a^\top \boldsymbol{\beta} + v_a$ of small areas for $a = 1, \dots, m$. Since θ involves the random effects v_i's, it is more appropriate to use the terminology of 'prediction of θ' than 'estimation of θ'.

In this section, we derive the best linear unbiased predictor of θ when $\boldsymbol{\psi}$ is known. Let us consider a class of linear and unbiased estimators $\boldsymbol{k}^\top \boldsymbol{y}$ for $\boldsymbol{k} \in \mathbb{R}^N$. Since $\mathrm{E}[\boldsymbol{k}^\top \boldsymbol{y}] = \boldsymbol{k}^\top \boldsymbol{X}\boldsymbol{\beta}$ and $\mathrm{E}[\theta] = \boldsymbol{c}_1^\top \boldsymbol{\beta}$, the vector $\boldsymbol{k}$ satisfies $\boldsymbol{k}^\top \boldsymbol{X} = \boldsymbol{c}_1^\top$.

Theorem 2.1 (BLUP) *The best linear unbiased predictor (BLUP) of* θ *is*

$$\widetilde{\theta}^{\mathrm{BLUP}} = \widetilde{\theta}^{\mathrm{BLUP}}(\boldsymbol{\psi}) = \boldsymbol{k}_0^\top \boldsymbol{y} = \boldsymbol{c}_1^\top \widetilde{\boldsymbol{\beta}} + \boldsymbol{c}_2^\top \widetilde{\boldsymbol{v}}, \tag{2.11}$$

where $\boldsymbol{k}_0^\top = \boldsymbol{c}_1^\top (\boldsymbol{X}^\top \boldsymbol{\Sigma}^{-1}\boldsymbol{X})^{-1}\boldsymbol{X}^\top \boldsymbol{\Sigma}^{-1} + \boldsymbol{c}_2^\top \boldsymbol{R}_v \boldsymbol{Z}^\top \boldsymbol{\Sigma}^{-1}\{\boldsymbol{I}_N - \boldsymbol{X}(\boldsymbol{X}^\top \boldsymbol{\Sigma}^{-1}\boldsymbol{X})^{-1}\boldsymbol{X}^\top \boldsymbol{\Sigma}^{-1}\}$ *and*

$$\begin{aligned}\widetilde{\boldsymbol{\beta}} &= \widetilde{\boldsymbol{\beta}}(\boldsymbol{\psi}) = (\boldsymbol{X}^\top \boldsymbol{\Sigma}^{-1}\boldsymbol{X})^{-1}\boldsymbol{X}^\top \boldsymbol{\Sigma}^{-1}\boldsymbol{y},\\ \widetilde{\boldsymbol{v}} &= \widetilde{\boldsymbol{v}}(\boldsymbol{\psi}) = \boldsymbol{R}_v \boldsymbol{Z}^\top \boldsymbol{\Sigma}^{-1}(\boldsymbol{y} - \boldsymbol{X}\widetilde{\boldsymbol{\beta}}).\end{aligned} \tag{2.12}$$

This shows that the generalized least squares (GLS) estimator $\widetilde{\boldsymbol{\beta}}$ *is the best linear unbiased estimator (BLUE) of* $\boldsymbol{\beta}$ *and* $\widetilde{\boldsymbol{v}}$ *is the best linear unbiased predictor of* $\boldsymbol{v}$.

Proof Since $\boldsymbol{k}$ and $\boldsymbol{k}_0$ satisfy $\boldsymbol{k}^\top \boldsymbol{X} = \boldsymbol{k}_0^\top \boldsymbol{X} = \boldsymbol{c}_1^\top$, it is noted that $\boldsymbol{k}^\top \boldsymbol{y} - \theta = \boldsymbol{k}^\top(\boldsymbol{y} - \boldsymbol{X}\boldsymbol{\beta}) + \boldsymbol{c}_2^\top \boldsymbol{v}$. Then,

$$\begin{aligned}\mathrm{Var}(\boldsymbol{k}^\top \boldsymbol{y}) &= \mathrm{E}[\{\boldsymbol{k}^\top(\boldsymbol{y} - \boldsymbol{X}\boldsymbol{\beta}) + \boldsymbol{c}_2^\top \boldsymbol{v}\}^2]\\ &= \boldsymbol{k}^\top \mathbf{Cov}\,(\boldsymbol{y})\boldsymbol{k} + 2\boldsymbol{k}^\top \mathbf{Cov}\,(\boldsymbol{y}, \boldsymbol{v})\boldsymbol{b} + \boldsymbol{c}_2^\top \mathbf{Cov}\,(\boldsymbol{v})\boldsymbol{c}_2,\end{aligned}$$

where $\mathbf{Cov}\,(\boldsymbol{y}) = \mathrm{E}[(\boldsymbol{y} - \boldsymbol{X}\boldsymbol{\beta})(\boldsymbol{y} - \boldsymbol{X}\boldsymbol{\beta})^\top] = \boldsymbol{\Sigma}$, $\mathbf{Cov}\,(\boldsymbol{y}, \boldsymbol{v}) = \mathrm{E}[(\boldsymbol{y} - \boldsymbol{X}\boldsymbol{\beta})\boldsymbol{v}^\top] = \boldsymbol{Z}\boldsymbol{R}_v$ and $\mathbf{Cov}\,(\boldsymbol{v}) = \boldsymbol{R}_v$. Thus,

$$\mathrm{Var}(\boldsymbol{k}^\top \boldsymbol{y}) = (\boldsymbol{k} - \boldsymbol{\Sigma}^{-1}\boldsymbol{Z}\boldsymbol{R}_v\boldsymbol{c}_2)^\top \boldsymbol{\Sigma}(\boldsymbol{k} - \boldsymbol{\Sigma}^{-1}\boldsymbol{Z}\boldsymbol{R}_v\boldsymbol{c}_2) + \boldsymbol{c}_2^\top(\boldsymbol{R}_v - \boldsymbol{R}_v\boldsymbol{Z}^\top \boldsymbol{\Sigma}^{-1}\boldsymbol{Z}\boldsymbol{R}_v)\boldsymbol{c}_2.$$

It is here noted that $\boldsymbol{k}_0^\top \boldsymbol{\Sigma}(\boldsymbol{k} - \boldsymbol{k}_0) = \boldsymbol{c}_2^\top \boldsymbol{R}_v \boldsymbol{Z}^\top(\boldsymbol{k} - \boldsymbol{k}_0)$, which implies that $(\boldsymbol{k}_0 - \boldsymbol{\Sigma}^{-1}\boldsymbol{Z}\boldsymbol{R}_v\boldsymbol{c}_2)^\top \boldsymbol{\Sigma}(\boldsymbol{k} - \boldsymbol{k}_0) = 0$. Hence,

$$\mathrm{Var}(\boldsymbol{k}^\top \boldsymbol{y}) = \mathrm{Var}(\boldsymbol{k}_0^\top \boldsymbol{y}) + (\boldsymbol{k} - \boldsymbol{k}_0)^\top \boldsymbol{\Sigma}(\boldsymbol{k} - \boldsymbol{k}_0),$$

which shows $\mathrm{Var}(\boldsymbol{k}^\top \boldsymbol{y}) \geq \mathrm{Var}(\boldsymbol{k}_0^\top \boldsymbol{y})$. When $\boldsymbol{c}_2 = \boldsymbol{0}$, this implies that $\widetilde{\boldsymbol{\beta}}$ is the best linear unbiased estimator (BLUE) of $\boldsymbol{\beta}$. It is also shown that $\widetilde{\boldsymbol{v}}$ is the best linear unbiased predictor of $\boldsymbol{v}$ when $\boldsymbol{c}_1 = \boldsymbol{0}$. □

There are another method for derivation of the BLUP under the normality of $\boldsymbol{v}$ and $\boldsymbol{\epsilon}$ given in (2.3). The joint probability density function of $(\boldsymbol{y}, \boldsymbol{v})$ is written as $(2\pi)^{-N/2}|\boldsymbol{R}_v|^{-1/2}|\boldsymbol{R}_e|^{-1/2} \cdot \exp\{-h(\boldsymbol{\beta}, \boldsymbol{v})/2\}$, where $h(\boldsymbol{\beta}, \boldsymbol{v}) = \boldsymbol{v}'\boldsymbol{R}_v^{-1}\boldsymbol{v} + (\boldsymbol{y} - \boldsymbol{X}\boldsymbol{\beta} - \boldsymbol{Z}\boldsymbol{v})^\top \boldsymbol{R}_e^{-1}(\boldsymbol{y} - \boldsymbol{X}\boldsymbol{\beta} - \boldsymbol{Z}\boldsymbol{v})$. The maximization of the joint pdf with respect to $\boldsymbol{\beta}$ and $\boldsymbol{v}$ is equivalent to the minimization of $h(\boldsymbol{\beta}, \boldsymbol{v})$, and the partial derivatives are

$$\begin{aligned}\frac{\partial h(\boldsymbol{\beta}, \boldsymbol{v})}{\partial \boldsymbol{\beta}} &= -2\boldsymbol{X}^\top \boldsymbol{R}_e^{-1}(\boldsymbol{y} - \boldsymbol{X}\boldsymbol{\beta} - \boldsymbol{Z}\boldsymbol{v}),\\ \frac{\partial h(\boldsymbol{\beta}, \boldsymbol{v})}{\partial \boldsymbol{v}} &= 2\boldsymbol{R}_v^{-1}\boldsymbol{v} - 2\boldsymbol{Z}^\top \boldsymbol{R}_e^{-1}(\boldsymbol{y} - \boldsymbol{X}\boldsymbol{\beta} - \boldsymbol{Z}\boldsymbol{v}).\end{aligned}$$

The matrix expression of $\partial h(\boldsymbol{\beta}, \boldsymbol{v})/\partial \boldsymbol{\beta} = \boldsymbol{0}$ and $\partial h(\boldsymbol{\beta}, \boldsymbol{v})/\partial \boldsymbol{v} = \boldsymbol{0}$ leads to the so-called *Mixed Model Equation* given by

$$\begin{pmatrix} X^\top R_e^{-1}X & X^\top R_e^{-1}Z \\ Z^\top R_e^{-1}X & Z^\top R_e^{-1}Z + R_v^{-1} \end{pmatrix} \begin{pmatrix} \widehat{\beta} \\ \widetilde{v} \end{pmatrix} = \begin{pmatrix} X^\top R_e^{-1}y \\ Z^\top R_e^{-1}y \end{pmatrix}. \tag{2.13}$$

This was provided by Henderson (1950), who showed the solutions are $\widetilde{\beta}$ and $\widetilde{v}$ given in (2.12).

Theorem 2.2 *The solutions of the mixed model equation in (2.13) are $\widetilde{\beta}$ and $\widetilde{v}$ given in (2.12).*

Proof The second equation in (2.13) is written as $Z^\top R_e^{-1}X\widehat{\beta} + (Z^\top R_e^{-1}Z + R_v^{-1})\widetilde{v} = Z^\top R_e^{-1}y$, which implies that

$$\widetilde{v} = (Z^\top R_e^{-1}Z + R_v^{-1})^{-1}Z^\top R_e^{-1}(y - X\beta). \tag{2.14}$$

It is noted that

$$\begin{aligned}
&(Z^\top R_e^{-1}Z + R_v^{-1})^{-1}Z^\top R_e^{-1} \\
&= R_v Z^\top R_e^{-1} - R_v \left\{(Z^\top R_e^{-1}Z + R_v^{-1}) - R_v^{-1}\right\}(Z^\top R_e^{-1}Z + R_v^{-1})^{-1}Z^\top R_e^{-1} \\
&= R_v Z^\top R_e^{-1} - R_v Z^\top R_e^{-1}Z(Z^\top R_e^{-1}Z + R_v^{-1})^{-1}Z^\top R_e^{-1} \\
&= R_v Z^\top \left\{R_e^{-1} - R_e^{-1}Z(R_v^{-1} + Z^\top R_e^{-1}Z)^{-1}Z^\top R_e^{-1}\right\} \\
&= R_v Z^\top \Sigma^{-1},
\end{aligned}$$

where at the last equality, we used the useful equality

$$\Sigma^{-1} = (ZR_vZ^\top + R_e)^{-1} = R_e^{-1} - R_e^{-1}Z(R_v^{-1} + Z^\top R_e^{-1}Z)^{-1}Z^\top R_e^{-1}. \tag{2.15}$$

Thus, $\widetilde{v}$ given in (2.14) is expressed as (2.12).

The first equation in (2.13) is $X^\top R_e^{-1}X\widehat{\beta} + X^\top R_e^{-1}Z\widetilde{v} = X^\top R_e^{-1}y$. Substituting $\widetilde{v} = R_vZ^\top\Sigma^{-1}(y - X\widehat{\beta})$ into the first equation yields $X^\top R_e^{-1}X\widehat{\beta} + X^\top R_e^{-1}ZR_vZ^\top\Sigma^{-1}(y - X\widehat{\beta}) = X^\top R_e^{-1}y$, or

$$X^\top R_e^{-1}(\Sigma - ZR_vZ^\top)\Sigma^{-1}X\widehat{\beta} = X^\top R_e^{-1}(\Sigma - ZR_vZ^\top)\Sigma^{-1}y.$$

Since $\Sigma = ZR_vZ^\top + R_e$, it is noted that $R_e^{-1}(\Sigma - ZR_vZ^\top) = I$. Thus, one gets the equation $X^\top\Sigma^{-1}X\widehat{\beta} = X^\top\Sigma^{-1}y$, which means that the solution $\widehat{\beta}$ is the GLS in (2.12). □

The BLUP of v is interpreted as an empirical Bayes estimator of v. The Bayes model in (2.3) with $\mu = X\beta + v$, the posterior distribution of v given y is

$$v|y \sim \mathrm{N}\left(R_vZ^\top\Sigma^{-1}(y - X\beta), R_v - R_vZ^\top\Sigma^{-1}ZR_v\right). \tag{2.16}$$

The Bayes estimator of $\boldsymbol{v}$ is $\widehat{\boldsymbol{v}}^{\mathrm{B}}(\boldsymbol{\beta}) = \boldsymbol{R}_v \boldsymbol{Z}^\top \boldsymbol{\Sigma}^{-1}(\boldsymbol{y} - \boldsymbol{X}\boldsymbol{\beta})$. The parameter $\boldsymbol{\beta}$ is estimated from the marginal distribution of $\boldsymbol{y}$, given $\mathrm{N}(\boldsymbol{X}\boldsymbol{\beta}, \boldsymbol{\Sigma})$, and the maximum likelihood estimator of $\boldsymbol{\beta}$ is the GLS $\widetilde{\boldsymbol{\beta}}$. Substituting $\widetilde{\boldsymbol{\beta}}$ into $\widehat{\boldsymbol{v}}^{\mathrm{B}}(\boldsymbol{\beta})$ yields the empirical Bayes estimator $\widehat{\boldsymbol{v}}^{\mathrm{EB}} = \widehat{\boldsymbol{v}}^{\mathrm{B}}(\widetilde{\boldsymbol{\beta}}) = \boldsymbol{R}_v \boldsymbol{Z}^\top \boldsymbol{\Sigma}^{-1}(\boldsymbol{y} - \boldsymbol{X}\widetilde{\boldsymbol{\beta}})$. This shows that the empirical Bayes estimator of $\boldsymbol{v}$ is given by $\widetilde{\boldsymbol{v}}$ in (2.12).

The distinction of the mixed model equation and the empirical Bayes estimation is that the former method is based on the mode of $\boldsymbol{v}$ in the joint density and the latter is based on the mean of the posterior distribution of $\boldsymbol{v}$. Although both methods give the same solution in normal distributions, their solutions are different in general. In the context of Bayesian statistics, the former method is called the *Maximum Bayesian Likelihood* method.

It is noted that the unobservable variable $\boldsymbol{v}$ can be predicted based on $\boldsymbol{y}$ when $\boldsymbol{y}$ is correlated with $\boldsymbol{v}$. In fact, the covariance matrix of $\boldsymbol{y}$ and $\boldsymbol{v}$ is

$$\mathbf{Cov}\begin{pmatrix} \boldsymbol{y} \\ \boldsymbol{v} \end{pmatrix} = \begin{pmatrix} \boldsymbol{\Sigma} & \boldsymbol{Z}\boldsymbol{R}_v \\ \boldsymbol{R}_v \boldsymbol{Z}' & \boldsymbol{R}_v \end{pmatrix},$$

which gives the conditional expectation $\mathrm{E}[\boldsymbol{v} \mid \boldsymbol{y}] = \boldsymbol{R}_v \boldsymbol{Z}^\top \boldsymbol{\Sigma}^{-1}(\boldsymbol{y} - \boldsymbol{X}\boldsymbol{\beta})$ under the normality. This consideration has been widely used in various fields like finite population models and incomplete data problems.

2.3 REML and General Estimating Equations

In the linear mixed models in (2.1), the covariance matrices $\boldsymbol{R}_v$ and $\boldsymbol{R}_e$ are, in general, functions of unknown parameters $\boldsymbol{\psi} = (\psi_1, \dots, \psi_q)^\top$ such as variance components, and we need to estimate them. Estimation of variance components has a long history, and various methods have been suggested in the literature. For example, the analysis of variance estimation (ANOVA), the minimum norm quadratic unbiased estimation (MINQUE), the maximum likelihood estimation (ML), and the restricted maximum likelihood estimation (REML) are well-known methods. See Rao and Kleffe (1988) and Searle et al. (1992) for the details.

The typical methods for estimating $\boldsymbol{\psi}$ are the *Maximum Likelihood* (ML) and *Restricted Maximum Likelihood* (REML). Substituting the GLS $\widetilde{\boldsymbol{\beta}} = \widetilde{\boldsymbol{\beta}}(\boldsymbol{\psi})$ into the marginal density function whose distribution is $\mathrm{N}(\boldsymbol{X}\boldsymbol{\beta}, \boldsymbol{\Sigma})$ for $\boldsymbol{\Sigma} = \boldsymbol{R}_e(\boldsymbol{\psi}) + \boldsymbol{Z}\boldsymbol{R}_v(\boldsymbol{\psi})\boldsymbol{Z}$, we can see that the ML estimator of $\boldsymbol{\psi}$ is the solution of minimizing the function $\log|\boldsymbol{\Sigma}| + (\boldsymbol{y} - \boldsymbol{X}\widetilde{\boldsymbol{\beta}})^\top \boldsymbol{\Sigma}^{-1}(\boldsymbol{y} - \boldsymbol{X}\widetilde{\boldsymbol{\beta}})$. On the other hand, let $\boldsymbol{K}$ be an $N \times (N - p)$ matrix satisfying $\boldsymbol{K}^\top \boldsymbol{X} = \boldsymbol{0}$. Then $\boldsymbol{K}^\top \boldsymbol{y} \sim \mathrm{N}(\boldsymbol{0}, \boldsymbol{K}^\top \boldsymbol{\Sigma} \boldsymbol{K})$, and the REML estimator is the solution of minimizing the function $\log|\boldsymbol{K}^\top \boldsymbol{\Sigma} \boldsymbol{K}| + \boldsymbol{y}^\top \boldsymbol{K}(\boldsymbol{K}^\top \boldsymbol{\Sigma} \boldsymbol{K})^{-1} \boldsymbol{K}^\top \boldsymbol{y}$. Let

$$\widetilde{\boldsymbol{P}} = \widetilde{\boldsymbol{P}}(\boldsymbol{\psi}) = \boldsymbol{\Sigma}^{-1} - \boldsymbol{\Sigma}^{-1}\boldsymbol{X}(\boldsymbol{X}^\top \boldsymbol{\Sigma}^{-1}\boldsymbol{X})^{-1}\boldsymbol{X}^\top \boldsymbol{\Sigma}^{-1}, \tag{2.17}$$

and note that $(\boldsymbol{y}-\boldsymbol{X}\widetilde{\boldsymbol{\beta}})^\top\boldsymbol{\Sigma}^{-1}(\boldsymbol{y}-\boldsymbol{X}\widetilde{\boldsymbol{\beta}})=\boldsymbol{y}^\top\widetilde{\boldsymbol{P}}\boldsymbol{y}$ and $\widetilde{\boldsymbol{P}}=\boldsymbol{K}(\boldsymbol{K}^\top\boldsymbol{\Sigma}\boldsymbol{K})^{-1}\boldsymbol{K}^\top$. Also note that $\partial_a\log|\boldsymbol{\Sigma}|=\mathrm{tr}\,(\boldsymbol{\Sigma}^{-1}\boldsymbol{\Sigma}_{(a)})$, $\partial_a\widetilde{\boldsymbol{P}}=-\widetilde{\boldsymbol{P}}\boldsymbol{\Sigma}_{(a)}\widetilde{\boldsymbol{P}}$, $\partial_a\log|\boldsymbol{K}^\top\boldsymbol{\Sigma}\boldsymbol{K}|=\mathrm{tr}\,[\widetilde{\boldsymbol{P}}\boldsymbol{\Sigma}_{(a)}]$ where $\partial_a=\partial/\partial\psi_a$ and $\boldsymbol{\Sigma}_{(a)}=\partial_a\boldsymbol{\Sigma}$. Thus, the ML and REML estimators are solutions of the following equations:

$$\text{[ML]}\quad (\boldsymbol{y}-\boldsymbol{X}\widetilde{\boldsymbol{\beta}})^\top\boldsymbol{\Sigma}^{-1}\boldsymbol{\Sigma}_{(a)}\boldsymbol{\Sigma}^{-1}(\boldsymbol{y}-\boldsymbol{X}\widetilde{\boldsymbol{\beta}})=\mathrm{tr}\,(\boldsymbol{\Sigma}^{-1}\boldsymbol{\Sigma}_{(a)}), \tag{2.18}$$

$$\text{[REML]}\quad (\boldsymbol{y}-\boldsymbol{X}\widetilde{\boldsymbol{\beta}})^\top\boldsymbol{\Sigma}^{-1}\boldsymbol{\Sigma}_{(a)}\boldsymbol{\Sigma}^{-1}(\boldsymbol{y}-\boldsymbol{X}\widetilde{\boldsymbol{\beta}})=\mathrm{tr}\,(\widetilde{\boldsymbol{P}}\boldsymbol{\Sigma}_{(a)}), \tag{2.19}$$

for $a=1,\ldots,q$. For discussions about which is better, ML or REML, see Sect. 6.10 in McCulloch and Searle (2001). Under the normality, Kubokawa (2011) showed that in estimation of variance components, REML is second-order unbiased, but ML has a second-order bias, while both have the same asymptotic covariance matrix. This suggests that REML is better than ML.

Although REML is derived under the normality, Eq. (2.19) can provide the consistent estimators without assuming the normality. We here suggest the general equations for estimating $\boldsymbol{\psi}$ without assuming the normality. Let $\boldsymbol{L}\boldsymbol{y}$ be a linear unbiased estimator of $\boldsymbol{\beta}$, where $\boldsymbol{L}=\boldsymbol{L}(\boldsymbol{\psi})$ is a $p\times N$ matrix of functions of $\boldsymbol{\psi}$ and satisfies $\boldsymbol{L}\boldsymbol{X}=\boldsymbol{I}$. Let $\boldsymbol{W}_a=\boldsymbol{W}_a(\boldsymbol{\psi})$ be an $N\times N$ matrix of functions of $\boldsymbol{\psi}$ for $a=1,\ldots,q$. The expectation $\mathrm{E}[(\boldsymbol{y}-\boldsymbol{X}\boldsymbol{L}\boldsymbol{y})^\top\boldsymbol{W}_a(\boldsymbol{y}-\boldsymbol{X}\boldsymbol{L}\boldsymbol{y})]$ is $\mathrm{tr}\,\{(\boldsymbol{I}-\boldsymbol{X}\boldsymbol{L})^\top\boldsymbol{W}_a(\boldsymbol{I}-\boldsymbol{X}\boldsymbol{L})\boldsymbol{\Sigma}\}$, which gives the general estimating equations

$$\boldsymbol{y}^\top(\boldsymbol{I}-\boldsymbol{X}\boldsymbol{L})^\top\boldsymbol{W}_a(\boldsymbol{I}-\boldsymbol{X}\boldsymbol{L})\boldsymbol{y}-\mathrm{tr}\,\{(\boldsymbol{I}-\boldsymbol{X}\boldsymbol{L})^\top\boldsymbol{W}_a(\boldsymbol{I}-\boldsymbol{X}\boldsymbol{L})\boldsymbol{\Sigma}\}=0, \tag{2.20}$$

for $a=1,\ldots,q$. For example, the choice of $\boldsymbol{W}_a=\boldsymbol{\Sigma}^{-1}\boldsymbol{\Sigma}_{(a)}\boldsymbol{\Sigma}^{-1}$ leads to the REML estimation, and other choices of $\boldsymbol{W}_a$ lead to different estimators of $\boldsymbol{\psi}$.

2.4 Asymptotic Properties

We now provide asymptotic properties of the general estimator $\widehat{\boldsymbol{\psi}}$ as the solution of (2.20). All the proofs of the theorems in this section are given in Sect. 2.5. Assume that the fourth moments are described as

$$\begin{aligned}\mathrm{E}[\{(\boldsymbol{R}_e^{-1/2}\boldsymbol{\epsilon})_i\}^4]&=\kappa_e+3,\\ \mathrm{E}[\{(\boldsymbol{R}_v^{-1/2}\boldsymbol{v})_i\}^4]&=\kappa_v+3,\end{aligned} \tag{2.21}$$

where $(\boldsymbol{a})_i$ is the i-th element of vector $\boldsymbol{a}$ and $\boldsymbol{A}^{1/2}$ is the symmetric root matrix of matrix $\boldsymbol{A}$. For $a=1,\ldots,k$, the estimating Eq. (2.20) is expressed as

$$\ell_a=\ell_a(\boldsymbol{\psi})=\boldsymbol{y}^\top\boldsymbol{C}_a\boldsymbol{y}-\mathrm{tr}\,(\boldsymbol{D}_a), \tag{2.22}$$

for

$$\begin{aligned} \boldsymbol{C}_a &= \boldsymbol{C}_a(\boldsymbol{\psi}) = (\boldsymbol{I} - \boldsymbol{X}\boldsymbol{L})^\top \boldsymbol{W}_a(\boldsymbol{I} - \boldsymbol{X}\boldsymbol{L}), \\ \boldsymbol{D}_a &= \boldsymbol{D}_a(\boldsymbol{\psi}) = \boldsymbol{C}_a\boldsymbol{\Sigma}. \end{aligned} \tag{2.23}$$

To establish the consistency of $\widehat{\boldsymbol{\psi}}$, we assume the following condition.

(C1) For $a = 1, \ldots, k$, $\ell_a(\boldsymbol{\psi})$ is a continuous function of $\boldsymbol{\psi}$ and has exactly one zero at $\widehat{\boldsymbol{\psi}}$. For any $\boldsymbol{\psi}$ and $\boldsymbol{\psi}'$, $\mathrm{E}_{\boldsymbol{\psi}}[\ell_a(\boldsymbol{\psi}')]$ satisfies $\mathrm{E}_{\boldsymbol{\psi}}[\ell_a(\boldsymbol{\psi})] = 0$ at $\boldsymbol{\psi}' = \boldsymbol{\psi}$ and $\mathrm{tr}\,[\{\boldsymbol{C}_a(\boldsymbol{\psi}')\boldsymbol{\Sigma}(\boldsymbol{\psi})\}^2] = O(N)$. Replacing the a-th element ψ_a with $\psi_a - \varepsilon$ for $\varepsilon > 0$, we use the notation $(\psi_a - \varepsilon, \boldsymbol{\psi}_{-a})$ for $(\psi_1, \ldots, \psi_{a-1}, \psi_a - \varepsilon, \psi_{a+1}, \ldots, \psi_q)^\top$. Then, it is assumed that $\mathrm{E}_{\boldsymbol{\psi}}[\ell_a(\psi_a - \varepsilon, \boldsymbol{\psi}_{-a})] < 0$ and $\mathrm{E}_{\boldsymbol{\psi}}[\ell_a(\psi_a + \varepsilon, \boldsymbol{\psi}_{-a})] > 0$ for $a = 1, \ldots, k$.

Theorem 2.3 *Assume that there exist the fourth moments* $\mathrm{E}[\|\boldsymbol{y}\|^4] < \infty$. *Under the condition* (C1), *the estimator* $\widehat{\boldsymbol{\psi}}$ *is consistent.*

We next show the $\sqrt{N}$-convergence of $\widehat{\boldsymbol{\psi}}$. It is noted that the Taylor series expansions of $\ell_a(\widehat{\boldsymbol{\psi}})$ around $\boldsymbol{\psi}$ are

$$\mathbf{0} = \mathbf{col}_a(\ell_a) + \mathbf{mat}_{ab}(\ell^\dagger_{a(b)})(\widehat{\boldsymbol{\psi}} - \boldsymbol{\psi}), \tag{2.24}$$

$$\begin{aligned} \mathbf{0} =& \mathbf{col}_a(\ell_a) + \mathbf{mat}_{ab}(\ell_{a(b)})(\widehat{\boldsymbol{\psi}} - \boldsymbol{\psi}) \\ &+ \frac{1}{2}\mathbf{col}_a\Big\{(\widehat{\boldsymbol{\psi}} - \boldsymbol{\psi})^\top \mathbf{mat}_{bc}(\ell^*_{a(bc)})(\widehat{\boldsymbol{\psi}} - \boldsymbol{\psi})\Big\}, \end{aligned} \tag{2.25}$$

for $\ell_a = \ell_a(\boldsymbol{\psi})$, $\ell_{a(b)} = \ell_{a(b)}(\boldsymbol{\psi})$, $\ell^\dagger_{a(b)} = \ell_{a(b)}(\boldsymbol{\psi}^\dagger)$, $\ell^*_{a(bc)} = \ell_{a(bc)}(\boldsymbol{\psi}^*)$, where $\boldsymbol{\psi}^\dagger$ and $\boldsymbol{\psi}^*$ are vectors on the line segment between $\boldsymbol{\psi}$ and $\widehat{\boldsymbol{\psi}}$, and the notations $\mathbf{col}_a(x_a)$ and $\mathbf{mat}_{ab}(x_{ab})$ are defined in (2.9). It can be seen that $\mathrm{E}[\ell_a] = 0$ and $\mathrm{E}[\ell_{a(b)}] = \mathrm{tr}\,(\boldsymbol{C}_{a(b)}\boldsymbol{\Sigma}) - \mathrm{tr}\,(\boldsymbol{D}_{a(b)}) = -\mathrm{tr}\,(\boldsymbol{C}_a\boldsymbol{\Sigma}_{(b)})$, because $\ell_{a(b)} = \boldsymbol{y}^\top\boldsymbol{C}_{a(b)}\boldsymbol{y} - \mathrm{tr}\,(\boldsymbol{D}_{a(b)})$ and $\mathrm{tr}\,(\boldsymbol{D}_{a(b)}) = \mathrm{tr}\,(\boldsymbol{C}_{a(b)}\boldsymbol{\Sigma}) + \mathrm{tr}\,(\boldsymbol{C}_a\boldsymbol{\Sigma}_{(b)})$. Using Lemma 2.1, we have the covariance

$$\mathrm{Cov}(\ell_a, \ell_b) = 2\mathrm{tr}\,(\boldsymbol{C}_a\boldsymbol{\Sigma}\boldsymbol{C}_b\boldsymbol{\Sigma}) + \kappa(\boldsymbol{C}_a, \boldsymbol{C}_b), \tag{2.26}$$

where

$$\begin{aligned} \kappa(\boldsymbol{C}, \boldsymbol{D}) =& \kappa_e \sum_{j=1}^N (\boldsymbol{R}_e^{1/2}\boldsymbol{C}\boldsymbol{R}_e^{1/2})_{jj} \cdot (\boldsymbol{R}_e^{1/2}\boldsymbol{D}\boldsymbol{R}_e^{1/2})_{jj} \\ &+ \kappa_v \sum_{j=1}^m (\boldsymbol{R}_v^{1/2}\boldsymbol{Z}^\top\boldsymbol{C}\boldsymbol{Z}\boldsymbol{R}_v^{1/2})_{jj} \cdot (\boldsymbol{R}_v^{1/2}\boldsymbol{Z}^\top\boldsymbol{D}\boldsymbol{Z}\boldsymbol{R}_v^{1/2})_{jj}. \end{aligned} \tag{2.27}$$

We assume the following condition.

(C2) For $a \in \{1, \ldots, q\}$, $\boldsymbol{C}_a(\boldsymbol{\psi})$ and $\boldsymbol{D}_a(\boldsymbol{\psi})$ are twice continuously differentiable. For $a, b, c \in \{1, \ldots, q\}$, the matrix $N^{-1}\mathbf{mat}_{ab}\{\mathrm{tr}\,(\boldsymbol{C}_a\boldsymbol{\Sigma}_{(b)})\}$ converges to a positive definite matrix, and $\mathrm{tr}\,(\boldsymbol{C}_{a(bc)}\boldsymbol{\Sigma}) - \mathrm{tr}\,(\boldsymbol{D}_{a(bc)})$, $2\mathrm{tr}\,(\boldsymbol{C}_a\boldsymbol{\Sigma}\boldsymbol{C}_b\boldsymbol{\Sigma}) + \kappa(\boldsymbol{C}_a, \boldsymbol{C}_b)$, $2\mathrm{tr}\,\{(\boldsymbol{C}_{a(b)}\boldsymbol{\Sigma})^2\} + \kappa(\boldsymbol{C}_{a(b)}, \boldsymbol{C}_{a(b)})$, and $2\mathrm{tr}\,\{(\boldsymbol{C}_{a(bc)}\boldsymbol{\Sigma})^2\} + \kappa(\boldsymbol{C}_{a(bc)}, \boldsymbol{C}_{a(bc)})$ are of order $O(N)$.

Theorem 2.4 *Under the conditions* (C1) *and* (C2), $\widehat{\boldsymbol{\psi}} - \boldsymbol{\psi} = O_p(N^{-1/2})$.

We derive the second-order bias and the asymptotic covariance matrix of $\widehat{\boldsymbol{\psi}}$. These quantities for some typical estimators were obtained by Datta and Lahiri (2000) under the normality. The following theorem was provided by Kubokawa et al. (2021) without assuming the normality. Define $\boldsymbol{A}$, $\boldsymbol{B}$, $\boldsymbol{E}_a$, and $\boldsymbol{F}_a$ by

$$\begin{aligned}
\boldsymbol{A} &= \mathbf{mat}_{ab}\{\mathrm{tr}\,(\boldsymbol{C}_a\boldsymbol{\Sigma}_{(b)})\},\\
\boldsymbol{B} &= \mathbf{mat}_{ab}\{2\mathrm{tr}\,(\boldsymbol{C}_a\boldsymbol{\Sigma}\boldsymbol{C}_b\boldsymbol{\Sigma}) + \kappa(\boldsymbol{C}_a, \boldsymbol{C}_b)\},\\
\boldsymbol{E}_a &= \mathbf{mat}_{bc}\{\mathrm{tr}\,(\boldsymbol{C}_{a(bc)}\boldsymbol{\Sigma}) - \mathrm{tr}\,(\boldsymbol{D}_{a(bc)})\},\\
\boldsymbol{F}_a &= \mathbf{mat}_{bc}\{2\mathrm{tr}\,(\boldsymbol{C}_{a(b)}\boldsymbol{\Sigma}\boldsymbol{C}_c\boldsymbol{\Sigma}) + \kappa(\boldsymbol{C}_{a(b)}, \boldsymbol{C}_c)\}.
\end{aligned} \tag{2.28}$$

We add a couple of assumptions.

(C3) For $a, b \in \{1, \ldots, q\}$, $\mathrm{E}[\{\ell^{\dagger}_{a(b)} + (\boldsymbol{A})_{ab}\}^2(\widehat{\psi}_b - \psi_b)^2] = o(N)$.

(C4) $2\mathrm{tr}\,(\boldsymbol{C}_{a(b)}\boldsymbol{\Sigma}\boldsymbol{C}_c\boldsymbol{\Sigma}) + \kappa(\boldsymbol{C}_{a(b)}, \boldsymbol{C}_c) = O(N)$, and the expectation of the remainder term $\mathrm{E}[\boldsymbol{r}(\boldsymbol{y})]$ is of order $o(N^{-1})$, where

$$\begin{aligned}
\boldsymbol{r}(\boldsymbol{y}) = &\boldsymbol{A}^{-1}\{\mathbf{mat}_{ab}(\ell_{a(b)}) + \boldsymbol{A}\}\{\widehat{\boldsymbol{\psi}} - \boldsymbol{\psi} - \boldsymbol{A}^{-1}\mathbf{col}_c(\ell_c)\}\\
&+ \boldsymbol{A}^{-1}\mathbf{col}_a\{(\widehat{\boldsymbol{\psi}} - \boldsymbol{\psi} - \boldsymbol{A}^{-1}\mathbf{col}_b(\ell_b))^\top \boldsymbol{E}_a\boldsymbol{A}^{-1}\mathbf{col}_c(\ell_c)\}\\
&+ \boldsymbol{A}^{-1}\mathbf{col}_a\{(\widehat{\boldsymbol{\psi}} - \boldsymbol{\psi} - \boldsymbol{A}^{-1}\mathbf{col}_b(\ell_b))^\top \boldsymbol{E}_a(\widehat{\boldsymbol{\psi}} - \boldsymbol{\psi} - \boldsymbol{A}^{-1}\mathbf{col}_c(\ell_c))\}\\
&+ \frac{1}{2}\boldsymbol{A}^{-1}\mathbf{col}_a[(\widehat{\boldsymbol{\psi}} - \boldsymbol{\psi})^\top\{\mathbf{mat}_{bc}(\ell^*_{a(bc)}) - \boldsymbol{E}_a\}(\widehat{\boldsymbol{\psi}} - \boldsymbol{\psi})].
\end{aligned} \tag{2.29}$$

Theorem 2.5 *Assume that conditions* (C1) *and* (C2) *hold.*

(1) *When condition* (C3) *is added, the covariance matrix of* $\widehat{\boldsymbol{\psi}}$ *is approximated as*

$$\mathbf{Cov}\,(\widehat{\boldsymbol{\psi}}) = \boldsymbol{A}^{-1}\boldsymbol{B}\boldsymbol{A}^{-1} + o(N^{-1}). \tag{2.30}$$

(2) *When condition* (C4) *is added, the second-order bias of* $\widehat{\boldsymbol{\psi}}$ *is*

$$\mathrm{E}[\widehat{\boldsymbol{\psi}} - \boldsymbol{\psi}] = \boldsymbol{A}^{-1}\left\{\mathbf{col}_a\{\mathrm{tr}\,(\boldsymbol{F}_a\boldsymbol{A}^{-1})\} + \frac{1}{2}\mathbf{col}_a\{\mathrm{tr}\,(\boldsymbol{E}_a\boldsymbol{A}^{-1}\boldsymbol{B}\boldsymbol{A}^{-1})\}\right\} + o(N^{-1}). \tag{2.31}$$

The following proposition shows that the second-order bias and the asymptotic covariance matrix given in Theorem 2.5 do not depend on $\boldsymbol{L}$.

(C5) For $\boldsymbol{L}$ such that $\boldsymbol{L}\boldsymbol{X} = \boldsymbol{I}$, assume that $(\boldsymbol{L}^\top\boldsymbol{X}^\top\boldsymbol{W}_a\boldsymbol{X}\boldsymbol{L})_{ij} = O(N^{-1})$, $\boldsymbol{L}\boldsymbol{\Sigma}\boldsymbol{L}^\top = O(N^{-1})$, and $(\boldsymbol{X}\boldsymbol{L})_{ij} = O(N^{-1})$ as $N \to \infty$.

Proposition 2.1 *Under condition* (C5), $\boldsymbol{A}$, $\boldsymbol{B}$, $\boldsymbol{E}_a$, *and* $\boldsymbol{F}_a$ *are approximated as*

$$\begin{aligned}
\boldsymbol{A} &= \mathbf{mat}_{ab}\{\mathrm{tr}\,(\boldsymbol{W}_a \boldsymbol{\Sigma}_{(b)})\} + O(1), \\
\boldsymbol{B} &= \mathbf{mat}_{ab}\{2\mathrm{tr}\,(\boldsymbol{W}_a \boldsymbol{\Sigma} \boldsymbol{W}_b \boldsymbol{\Sigma}) + \kappa(\boldsymbol{W}_a, \boldsymbol{W}_b)\} + O(1), \\
\boldsymbol{E}_a &= \mathbf{mat}_{bc}\{\mathrm{tr}\,(\boldsymbol{W}_{a(bc)} \boldsymbol{\Sigma}) - \mathrm{tr}\,(\boldsymbol{D}_{a(bc)})\} + O(1) \qquad (2.32) \\
&= -\mathbf{mat}_{bc}\{\mathrm{tr}\,(\boldsymbol{W}_{a(b)} \boldsymbol{\Sigma}_{(c)}) + \mathrm{tr}\,(\boldsymbol{W}_{a(c)} \boldsymbol{\Sigma}_{(b)}) + \mathrm{tr}\,(\boldsymbol{W}_a \boldsymbol{\Sigma}_{(bc)})\}, \\
\boldsymbol{F}_a &= \mathbf{mat}_{bc}\{2\mathrm{tr}\,(\boldsymbol{W}_{a(b)} \boldsymbol{\Sigma} \boldsymbol{W}_c \boldsymbol{\Sigma}) + \kappa(\boldsymbol{W}_{a(b)}, \boldsymbol{W}_c)\} + O(1).
\end{aligned}$$

Two typical choices of $\boldsymbol{L}$ are $\boldsymbol{L}^{\mathrm{G}} = (\boldsymbol{X}^\top \boldsymbol{\Sigma}^{-1} \boldsymbol{X})^{-1} \boldsymbol{X}^\top \boldsymbol{\Sigma}^{-1}$ and $\boldsymbol{L}^{\mathrm{O}} = (\boldsymbol{X}^\top \boldsymbol{X})^{-1} \boldsymbol{X}^\top$, which correspond to the GLS estimator $\widetilde{\boldsymbol{\beta}}$ and ordinary least squares (OLS) estimator $\widehat{\boldsymbol{\beta}}^{\mathrm{O}}$. However, Theorem 2.5 and Proposition 2.1 tell us that the second-order bias and the asymptotic covariance matrix do not depend on such a choice of $\boldsymbol{L}$. This is an essential observation from Theorem 2.5 and Proposition 2.1, and the specific form of $\widehat{\boldsymbol{\beta}}^{\mathrm{L}}$ in the estimating Eq. (2.20) is irrelevant to the asymptotic properties of $\widehat{\boldsymbol{\psi}}$ as long as $\widehat{\boldsymbol{\beta}}^{\mathrm{L}}$ is unbiased. Hence, it would be better to use a simpler form of $\boldsymbol{L}\boldsymbol{y}$, namely, $\boldsymbol{L} = (\boldsymbol{X}^\top \boldsymbol{X})^{-1} \boldsymbol{X}^\top$, corresponding to the ordinary least squares estimators of $\boldsymbol{\beta}$. On the other hand, the choice of $\boldsymbol{W}_a$ affects the asymptotic properties.

The second-order unbiasedness is one of the desirable properties of estimators $\widehat{\boldsymbol{\psi}}$. From Theorem 2.5 and Proposition 2.1, we need to use $\boldsymbol{W}_a$ such that the leading term in (2.31) is $\boldsymbol{0}$ to achieve second-order unbiasedness of $\widehat{\boldsymbol{\psi}}$. In typical linear mixed models such as the Fay–Herriot and nested error regression models, the covariance matrix $\boldsymbol{\Sigma}$ is a linear function of $\boldsymbol{\psi}$. In this case, $\boldsymbol{\Sigma}_{(bc)} = \boldsymbol{0}$, which simplifies the condition for the second-order unbiasedness in (2.31). When $\kappa_e = \kappa_v = 0$, the estimator $\widehat{\boldsymbol{\psi}}$ is second-order unbiased if

$$\begin{aligned}
&\mathbf{mat}_{bc}\{\mathrm{tr}\,(\boldsymbol{W}_{a(b)} \boldsymbol{\Sigma} \boldsymbol{W}_c \boldsymbol{\Sigma})\} \qquad (2.33) \\
&= \mathbf{mat}_{bc}\{\mathrm{tr}\,(\boldsymbol{W}_{a(b)} \boldsymbol{\Sigma}_{(c)})\}\big[\mathbf{mat}_{ab}\{\mathrm{tr}\,(\boldsymbol{W}_a \boldsymbol{\Sigma}_{(b)})\}\big]^{-1} \mathbf{mat}_{bc}\{\mathrm{tr}\,(\boldsymbol{W}_b \boldsymbol{\Sigma} \boldsymbol{W}_c \boldsymbol{\Sigma})\},
\end{aligned}$$

for $a = 1, \ldots, q$. This condition is investigated below for specific choices of $\boldsymbol{W}_a$.

In what follows, we assume that $\boldsymbol{\Sigma}$ is a linear function of $\boldsymbol{\psi}$, which are satisfied in typical linear mixed models such as the Fay–Herriot and nested error regression models. We consider the three candidates for $\boldsymbol{W}_a$:

$$\boldsymbol{W}_a^{\mathrm{RE}} = \boldsymbol{\Sigma}^{-1} \boldsymbol{\Sigma}_{(a)} \boldsymbol{\Sigma}^{-1}, \quad \boldsymbol{W}_a^{\mathrm{FH}} = (\boldsymbol{\Sigma}^{-1} \boldsymbol{\Sigma}_{(a)} + \boldsymbol{\Sigma}_{(a)} \boldsymbol{\Sigma}^{-1})/2 \;\text{ and }\; \boldsymbol{W}_a^{\mathrm{Q}} = \boldsymbol{\Sigma}_{(a)},$$

which are motivated from the REML estimator, the Fay–Herriot moment estimator (Fay and Herriot 1979), and the Prasad–Rao unbiased estimator (Prasad and Rao 1990) under the Fay–Herriot model. The estimators induced from $\boldsymbol{W}_a^{\mathrm{RE}}$, $\boldsymbol{W}_a^{\mathrm{FH}}$, and $\boldsymbol{W}_a^{\mathrm{Q}}$ are called here the REML-type, FH-type, and PR-type estimators, respectively. From Theorem 2.5 and Proposition 2.1, we can derive the asymptotic properties of the three estimators. When $\boldsymbol{\Sigma}$ is a linear function of $\boldsymbol{\psi}$, the asymptotic variances and second-order biases are simplified in the case of $\kappa_e = \kappa_v = 0$, which is satisfied in the normal distributions.

Theorem 2.6 *Assume that conditions* (C1)–(C5) *hold and that* $\boldsymbol{\Sigma}$ *is a linear function of* $\boldsymbol{\psi}$. *Let* $\widehat{\boldsymbol{\psi}}^{\mathrm{RE}}$, $\widehat{\boldsymbol{\psi}}^{\mathrm{FH}}$, *and* $\widehat{\boldsymbol{\psi}}^{\mathrm{Q}}$ *be the estimators based on* $\mathbf{W}_a^{\mathrm{RE}}$, $\mathbf{W}_a^{\mathrm{FH}}$, *and* $\mathbf{W}_a^{\mathrm{Q}}$, *respectively. Under the condition* $\kappa_e = \kappa_v = 0$, *the following results hold.*

(a) *REML-type estimator* $\widehat{\boldsymbol{\psi}}^{\mathrm{RE}}$ *is second-order unbiased and has the asymptotic covariance matrix* $2\boldsymbol{A}_{\mathrm{RE}}^{-1}$, *where* $(\boldsymbol{A}_{\mathrm{RE}})_{ab} = \mathrm{tr}\,(\boldsymbol{\Sigma}^{-1}\boldsymbol{\Sigma}_{(a)}\boldsymbol{\Sigma}^{-1}\boldsymbol{\Sigma}_{(b)})$.

(b) *FH-type estimator* $\widehat{\boldsymbol{\psi}}^{\mathrm{FH}}$ *is not second-order unbiased. The asymptotic covariance matrix is*

$$\boldsymbol{A}_{\mathrm{FH}}^{-1}\boldsymbol{B}_{\mathrm{FH}}\boldsymbol{A}_{\mathrm{FH}}^{-1},$$

for $(\boldsymbol{A}_{\mathrm{FH}})_{ab} = \mathrm{tr}\,(\boldsymbol{\Sigma}^{-1}\boldsymbol{\Sigma}_{(a)}\boldsymbol{\Sigma}_{(b)})$ *and* $(\boldsymbol{B}_{\mathrm{FH}})_{ab} = \mathrm{tr}\,(\boldsymbol{\Sigma}_{(a)}\boldsymbol{\Sigma}_{(b)}) + \mathrm{tr}\,(\boldsymbol{\Sigma}^{-1}\boldsymbol{\Sigma}_{(a)}\boldsymbol{\Sigma}\boldsymbol{\Sigma}_{(b)})$. *The second-order bias is*

$$\boldsymbol{A}_{\mathrm{FH}}^{-1}\mathbf{col}_a\left\{\mathrm{tr}\,(\boldsymbol{F}_a\boldsymbol{A}_{\mathrm{FH}}^{-1}) + \frac{1}{2}\mathrm{tr}\,(\boldsymbol{E}_a\boldsymbol{A}_{\mathrm{FH}}^{-1}\boldsymbol{B}_{\mathrm{FH}}\boldsymbol{A}_{\mathrm{FH}}^{-1})\right\},$$

for $(\boldsymbol{E}_a)_{bc} = 2\mathrm{tr}\,\{\boldsymbol{\Sigma}_{(a)}\boldsymbol{\Sigma}_{(c)}\boldsymbol{\Sigma}^{-1}\boldsymbol{\Sigma}_{(b)}\boldsymbol{\Sigma}^{-1}\}$ *and*

$$(\boldsymbol{F}_a)_{bc} = -\mathrm{tr}\,\{\boldsymbol{\Sigma}_{(a)}\boldsymbol{\Sigma}^{-1}\boldsymbol{\Sigma}_{(b)}(\boldsymbol{\Sigma}_{(c)} + \boldsymbol{\Sigma}^{-1}\boldsymbol{\Sigma}_{(c)}\boldsymbol{\Sigma})\}.$$

(c) *PR-type estimator* $\widehat{\boldsymbol{\psi}}^{\mathrm{Q}}$ *is second-order unbiased. The asymptotic covariance matrix is* $\boldsymbol{A}_{\mathrm{Q}}^{-1}\boldsymbol{B}_{\mathrm{Q}}\boldsymbol{A}_{\mathrm{Q}}^{-1}$ *where* $(\boldsymbol{A}_{\mathrm{Q}})_{ab} = \mathrm{tr}\,(\boldsymbol{\Sigma}_{(a)}\boldsymbol{\Sigma}_{(b)})$ *and* $(\boldsymbol{B}_{\mathrm{Q}})_{ab} = 2\mathrm{tr}\,(\boldsymbol{\Sigma}_{(a)}\boldsymbol{\Sigma}\boldsymbol{\Sigma}_{(b)}\boldsymbol{\Sigma})$.

In Theorem 2.6, the linearity of $\boldsymbol{\Sigma}(\boldsymbol{\psi})$ on $\boldsymbol{\psi}$ is only used to compute the second-order bias. The expressions for the asymptotic covariances hold, in general, without such constraints. Without assuming $\kappa_e = \kappa_v = 0$, the estimator $\widehat{\boldsymbol{\psi}}^{\mathrm{RE}}$ has the second-order bias, while $\widehat{\boldsymbol{\psi}}^{\mathrm{Q}}$ remains second-order unbiased.

It is noted that the REML type is the most efficient in the normal distributions with $\kappa_e = \kappa_v = 0$. This implies that the following inequality holds for any $\boldsymbol{W}_a$:

$$\begin{aligned}[\mathbf{mat}_{a,b}\{\mathrm{tr}\,(\boldsymbol{W}_a\boldsymbol{\Sigma}_{(b)})\}]^{-1}\mathbf{mat}_{a,b}\{\mathrm{tr}\,(\boldsymbol{W}_a\boldsymbol{\Sigma}\boldsymbol{W}_b\boldsymbol{\Sigma})\}[\mathbf{mat}_{a,b}\{\mathrm{tr}\,(\boldsymbol{W}_a\boldsymbol{\Sigma}_{(b)})\}]^{-1}\\ \geq [\mathbf{mat}_{a,b}\{\mathrm{tr}\,(\boldsymbol{\Sigma}^{-1}\boldsymbol{\Sigma}_{(a)}\boldsymbol{\Sigma}^{-1}\boldsymbol{\Sigma}_{(b)})\}]^{-1}. \qquad (2.34)\end{aligned}$$

However, it should be remarked that REML is not necessarily efficient without assuming $\kappa_e = 0$ and $\kappa_v = 0$.

2.5 Proofs of the Asymptotic Results

We here provide the proofs of the asymptotic results given in Sect. 2.4. The following lemma is useful for the proofs of the theorems.

Lemma 2.1 *Let* $\boldsymbol{u} = \boldsymbol{\epsilon} + \boldsymbol{Z}\boldsymbol{v}$*. Then, for matrices* $\boldsymbol{C}$ *and* $\boldsymbol{D}$*, it holds that*

$$\mathrm{E}[\boldsymbol{u}^\top \boldsymbol{C}\boldsymbol{u}\boldsymbol{u}^\top \boldsymbol{D}\boldsymbol{u}] = 2\mathrm{tr}\,(\boldsymbol{C}\boldsymbol{\Sigma}\boldsymbol{D}\boldsymbol{\Sigma}) + \mathrm{tr}\,(\boldsymbol{C}\boldsymbol{\Sigma})\mathrm{tr}\,(\boldsymbol{D}\boldsymbol{\Sigma}) + \kappa(\boldsymbol{C}, \boldsymbol{D}), \tag{2.35}$$

where $\kappa(\boldsymbol{C}, \boldsymbol{D})$ *is given in (2.27).*

Proof It is demonstrated that

$$\begin{aligned}
&\mathrm{E}[\boldsymbol{u}^\top \boldsymbol{C}\boldsymbol{u}\boldsymbol{u}^\top \boldsymbol{D}\boldsymbol{u}]\\
=&\mathrm{E}[\boldsymbol{\epsilon}^\top \boldsymbol{C}\boldsymbol{\epsilon}\boldsymbol{\epsilon}^\top \boldsymbol{D}\boldsymbol{\epsilon}] + \mathrm{E}[\boldsymbol{v}^\top \boldsymbol{Z}^\top \boldsymbol{C}\boldsymbol{Z}\boldsymbol{v}\boldsymbol{v}^\top \boldsymbol{Z}^\top \boldsymbol{D}\boldsymbol{Z}\boldsymbol{v}] + \mathrm{tr}\,(\boldsymbol{C}\boldsymbol{R}_e)\mathrm{tr}\,(\boldsymbol{D}\boldsymbol{Z}\boldsymbol{R}_v\boldsymbol{Z}^\top)\\
&+ \mathrm{tr}\,(\boldsymbol{D}\boldsymbol{R}_e)\mathrm{tr}\,(\boldsymbol{C}\boldsymbol{Z}\boldsymbol{R}_v\boldsymbol{Z}^\top) + 4\mathrm{tr}\,(\boldsymbol{C}\boldsymbol{R}_e\boldsymbol{D}\boldsymbol{Z}\boldsymbol{R}_v\boldsymbol{Z}^\top).
\end{aligned}$$

Let $\boldsymbol{x} = (x_1, \dots, x_N)^\top = \boldsymbol{R}_e^{-1/2}\boldsymbol{\epsilon}$, $\widetilde{\boldsymbol{C}} = \boldsymbol{R}_e^{1/2}\boldsymbol{C}\boldsymbol{R}_e^{1/2}$, and $\widetilde{\boldsymbol{D}} = \boldsymbol{R}_e^{1/2}\boldsymbol{D}\boldsymbol{R}_e^{1/2}$. Then, $\mathrm{E}[\boldsymbol{x}] = \boldsymbol{0}$, $\mathrm{E}[\boldsymbol{x}\boldsymbol{x}^\top] = \boldsymbol{I}_N$, $\mathrm{E}[x_a^4] = \kappa_e + 3$, $a = 1, \dots, N$, and $\mathrm{E}[\boldsymbol{\epsilon}^\top \boldsymbol{C}\boldsymbol{\epsilon}\boldsymbol{\epsilon}^\top \boldsymbol{D}\boldsymbol{\epsilon}] = \mathrm{E}[\boldsymbol{x}^\top \widetilde{\boldsymbol{C}}\boldsymbol{x}\boldsymbol{x}^\top \widetilde{\boldsymbol{D}}\boldsymbol{x}]$. Let $\delta_{a=b=c=d} = 1$ for $a = b = c = d$, and, otherwise, $\delta_{a=b=c=d}=0$. The notation $\delta_{a=b\neq c=d}$ is defined similarly. It is observed that for $a, b, c, d = 1, \dots, N$,

$$\begin{aligned}
&\mathrm{E}[x_a(\widetilde{\boldsymbol{C}})_{ab}x_bx_c(\widetilde{\boldsymbol{D}})_{cd}x_d]\\
=&\mathrm{E}[x_a^4(\widetilde{\boldsymbol{C}})_{aa}(\widetilde{\boldsymbol{D}})_{aa}\delta_{a=b=c=d} + x_a^2x_c^2(\widetilde{\boldsymbol{C}})_{aa}(\widetilde{\boldsymbol{D}})_{cc}\delta_{a=b\neq c=d}\\
&+ 2x_a^2x_b^2(\widetilde{\boldsymbol{C}})_{ab}(\widetilde{\boldsymbol{D}})_{ab}\delta_{a=c\neq b=d}]\\
=&(\kappa_e + 3)(\widetilde{\boldsymbol{C}})_{aa}(\widetilde{\boldsymbol{D}})_{aa}\delta_{a=b=c=d} + (\widetilde{\boldsymbol{C}})_{aa}(\widetilde{\boldsymbol{D}})_{cc}\delta_{a=b\neq c=d} + 2(\widetilde{\boldsymbol{C}})_{ab}(\widetilde{\boldsymbol{D}})_{ab}\delta_{a=c\neq b=d}\\
=&\kappa_e(\widetilde{\boldsymbol{C}})_{aa}(\widetilde{\boldsymbol{D}})_{aa}\delta_{a=b=c=d} + (\widetilde{\boldsymbol{C}})_{aa}(\widetilde{\boldsymbol{D}})_{cc}\delta_{a=b}\delta_{c=d} + 2(\widetilde{\boldsymbol{C}})_{ab}(\widetilde{\boldsymbol{D}})_{ab}\delta_{a=c}\delta_{b=d},
\end{aligned}$$

which implies that

$$\begin{aligned}
&\sum_{a,b,c,d}\mathrm{E}[x_a(\widetilde{\boldsymbol{C}})_{ab}x_bx_c(\widetilde{\boldsymbol{D}})_{cd}x_d]\\
&=\kappa_e\sum_{a=1}^N(\widetilde{\boldsymbol{C}})_{aa}(\widetilde{\boldsymbol{D}})_{aa} + \sum_{a=1}^N(\widetilde{\boldsymbol{C}})_{aa}\sum_{c=1}^N(\widetilde{\boldsymbol{D}})_{cc} + 2\sum_{a=1}^N\sum_{b=1}^N(\widetilde{\boldsymbol{C}})_{ab}(\widetilde{\boldsymbol{D}})_{ab},
\end{aligned}$$

or $\mathrm{E}[\boldsymbol{\epsilon}^\top \boldsymbol{C}\boldsymbol{\epsilon}\boldsymbol{\epsilon}^\top \boldsymbol{D}\boldsymbol{\epsilon}]=2\mathrm{tr}\,(\boldsymbol{C}\boldsymbol{R}_e\boldsymbol{D}\boldsymbol{R}_e) + \mathrm{tr}\,(\boldsymbol{C}\boldsymbol{R}_e)\mathrm{tr}\,(\boldsymbol{D}\boldsymbol{R}_e) + \kappa_e\sum_{i=1}^N(\boldsymbol{R}_e^{1/2}\boldsymbol{C}\boldsymbol{R}_e^{1/2})_{ii}\cdot(\boldsymbol{R}_e^{1/2}\boldsymbol{D}\boldsymbol{R}_e^{1/2})_{ii}$. Similarly,

$$\begin{aligned}\mathrm{E}[\boldsymbol{v}^\top \boldsymbol{Z}^\top \boldsymbol{C}\boldsymbol{Z}\boldsymbol{v}\boldsymbol{v}^\top \boldsymbol{Z}^\top \boldsymbol{D}\boldsymbol{Z}\boldsymbol{v}] =& 2\mathrm{tr}\,(\boldsymbol{C}\boldsymbol{Z}\boldsymbol{R}_v\boldsymbol{Z}^\top \boldsymbol{D}\boldsymbol{Z}\boldsymbol{R}_v\boldsymbol{Z}^\top) + \mathrm{tr}\,(\boldsymbol{C}\boldsymbol{Z}\boldsymbol{R}_v\boldsymbol{Z}^\top)\mathrm{tr}\,(\boldsymbol{D}\boldsymbol{Z}\boldsymbol{R}_v\boldsymbol{Z}^\top)\\ &+ \kappa_v \sum_{i=1}^m (\boldsymbol{R}_v^{1/2}\boldsymbol{Z}^\top \boldsymbol{C}\boldsymbol{Z}\boldsymbol{R}_v^{1/2})_{ii} \cdot (\boldsymbol{R}_v^{1/2}\boldsymbol{Z}^\top \boldsymbol{D}\boldsymbol{Z}\boldsymbol{R}_v^{1/2})_{ii}.\end{aligned}$$

Thus, we have

$$\begin{aligned}\mathrm{E}[\boldsymbol{u}^\top \boldsymbol{C}\boldsymbol{u}\boldsymbol{u}^\top \boldsymbol{D}\boldsymbol{u}] =& 2\mathrm{tr}\,(\boldsymbol{C}\boldsymbol{R}_e\boldsymbol{D}\boldsymbol{R}_e) + \mathrm{tr}\,(\boldsymbol{C}\boldsymbol{R}_e)\mathrm{tr}\,(\boldsymbol{D}\boldsymbol{R}_e) + 2\mathrm{tr}\,(\boldsymbol{C}\boldsymbol{Z}\boldsymbol{R}_v\boldsymbol{Z}^\top \boldsymbol{D}\boldsymbol{Z}\boldsymbol{R}_v\boldsymbol{Z}^\top)\\ &+ \mathrm{tr}\,(\boldsymbol{C}\boldsymbol{Z}\boldsymbol{R}_v\boldsymbol{Z}^\top)\mathrm{tr}\,(\boldsymbol{D}\boldsymbol{Z}\boldsymbol{R}_v\boldsymbol{Z}^\top) + \mathrm{tr}\,(\boldsymbol{C}\boldsymbol{R}_e)\mathrm{tr}\,(\boldsymbol{D}\boldsymbol{Z}\boldsymbol{R}_v\boldsymbol{Z}^\top)\\ &+ \mathrm{tr}\,(\boldsymbol{D}\boldsymbol{R}_e)\mathrm{tr}\,(\boldsymbol{C}\boldsymbol{Z}\boldsymbol{R}_v\boldsymbol{Z}^\top) + 4\mathrm{tr}\,(\boldsymbol{C}\boldsymbol{R}_e\boldsymbol{D}\boldsymbol{Z}\boldsymbol{R}_v\boldsymbol{Z}^\top) + h_{(\boldsymbol{C},\boldsymbol{D})},\end{aligned}$$

which can be rewritten as the expression in (2.35) for $\boldsymbol{\Sigma} = \boldsymbol{R}_e + \boldsymbol{Z}\boldsymbol{R}_v\boldsymbol{Z}^\top$. □

***Proof of Theorem* 2.3** For the proof of consistency, we use the same arguments as in Lemma 5.10 of van der Vaart (1998). Let $\boldsymbol{\psi}$ be the true value of the parameter. For $\boldsymbol{\psi}'$, let $\eta_a^{(N)}(\boldsymbol{\psi}') = N^{-1}\ell_a(\boldsymbol{\psi}')$ and $\eta_a(\boldsymbol{\psi}') = \mathrm{E}_{\boldsymbol{\psi}}[\eta_a^{(N)}(\boldsymbol{\psi}')]$. Then $\eta_a(\boldsymbol{\psi}') = N^{-1}\mathrm{tr}\,\{\boldsymbol{C}_a(\boldsymbol{\psi}')\boldsymbol{\Sigma}(\boldsymbol{\psi})\} - N^{-1}\mathrm{tr}\,\{\boldsymbol{D}_a(\boldsymbol{\psi}')\}$. Using Lemma 2.1, we can see that

$$\begin{aligned}\mathrm{E}[\{\eta_a^{(N)}(\boldsymbol{\psi}') - \eta_a(\boldsymbol{\psi}')\}^2] =& \frac{2}{N^2}\mathrm{tr}\,[\{\boldsymbol{C}_a(\boldsymbol{\psi}')\boldsymbol{\Sigma}(\boldsymbol{\psi})\}^2]\\ &+ \frac{\kappa_e}{N^2}\sum_{i=1}^N (\boldsymbol{R}_e^{1/2}\boldsymbol{C}_a(\boldsymbol{\psi}')\boldsymbol{R}_e^{1/2})_{ii}^2 + \frac{\kappa_v}{N^2}\sum_{i=1}^m (\boldsymbol{R}_v^{1/2}\boldsymbol{Z}^\top \boldsymbol{C}_a(\boldsymbol{\psi}')\boldsymbol{Z}\boldsymbol{R}_v^{1/2})_{ii}^2,\end{aligned}$$

where $\boldsymbol{R}_e = \boldsymbol{R}_e(\boldsymbol{\psi})$ and $\boldsymbol{R}_v = \boldsymbol{R}_v(\boldsymbol{\psi})$. From condition (C1), we have $\mathrm{E}[\{\eta_a^{(N)}(\boldsymbol{\psi}') - \eta_a(\boldsymbol{\psi}')\}^2] = O(N^{-1})$, namely, $\eta_a^{(N)}(\boldsymbol{\psi}')$ converges in probability to $\eta_a(\boldsymbol{\psi}')$.

Since $\boldsymbol{\psi}$ is the true value, it is noted that $\eta_a(\boldsymbol{\psi}) = 0$. From condition (C1), we have $\eta_a(\psi_a - \varepsilon, \boldsymbol{\psi}_{-a}) < 0$ and $\eta_a(\psi_a + \varepsilon, \boldsymbol{\psi}_{-a}) > 0$.

Let $E_a^{(-)} = \{\boldsymbol{y} \mid \eta_a^{(N)}(\psi_a - \varepsilon, \boldsymbol{\psi}_{-a}) < 0\}$ and $E_a^{(+)} = \{\boldsymbol{y} \mid \eta_a^{(N)}(\psi_a + \varepsilon, \boldsymbol{\psi}_{-a}) > 0\}$. From (C1), $\eta_a^{(N)}(\boldsymbol{\psi}')$ is continuous and has exactly one zero at $\widehat{\boldsymbol{\psi}}$. Thus, $E_a^{(-)} \cap E_a^{(+)} \subset \{\boldsymbol{y} \mid \psi_a - \varepsilon < \widehat{\psi}_a < \psi_a + \varepsilon\}$, so that $\cap_{a=1}^q (E_a^{(-)} \cap E_a^{(+)}) \subset \{\boldsymbol{y} \mid \psi_a - \varepsilon < \widehat{\psi}_a < \psi_a + \varepsilon, a = 1, \ldots, q\}$. From the Bonferroni inequality, it follows that

$$\begin{aligned}&\mathrm{P}(\psi_a - \varepsilon < \widehat{\psi}_a < \psi_a + \varepsilon, a = 1, \ldots, q)\\ \geq& \mathrm{P}(\cap_{a=1}^q (E_a^{(-)} \cap E_a^{(+)})) \geq 1 - \sum_{a=1}^q \{\mathrm{P}((E_a^{(-)})^c) + \mathrm{P}((E_a^{(+)})^c)\}.\end{aligned}$$

Since $\eta_a^{(N)}(\boldsymbol{\psi}')$ converges in probability to $\eta_a(\boldsymbol{\psi}')$, we have $\mathrm{P}((E_a^{(-)})^c) = 1 - \mathrm{P}(E_a^{(-)})$ and

$$\mathrm{P}(\eta_a^{(N)}(\psi_a - \varepsilon, \boldsymbol{\psi}_{-a}) < 0) \to 1.$$

Hence, $\mathrm{P}(\psi_a - \varepsilon < \widehat{\psi}_a < \psi_a + \varepsilon, a = 1, \ldots, q) \to 1$. □

Proof of Theorem 2.4 From the expansion (2.25), it follows that

$$\mathbf{col}_a(N^{-1}\ell_a) = -\Big[\mathbf{mat}_{ab}(N^{-1}\ell_{a(b)}) + \frac{1}{2}\boldsymbol{G}\Big](\widehat{\boldsymbol{\psi}} - \boldsymbol{\psi}),$$

where $\boldsymbol{G} = (g_{ac})$ is a $q \times q$ matrix with the (a, c)-element

$$g_{ac} = \sum_{b=1}^{q}(\widehat{\psi}_b - \psi_b)N^{-1}\ell_{a(bc)}(\boldsymbol{\psi}^*).$$

It can be seen that $N^{-1}\ell_a = O_p(N^{-1/2})$ under condition (C1). It is shown that $N^{-1}\ell_{a(b)} = O_p(1)$ under condition (C2), because $\mathrm{Var}(\ell_{a(b)}) = 2\mathrm{tr}\,\{(\boldsymbol{C}_{a(b)}\boldsymbol{\Sigma})^2\} + \kappa(\boldsymbol{C}_{a(b)}, \boldsymbol{C}_{a(b)})$. Consider a compact ball B with $\boldsymbol{\psi} \in B$. Note that $\mathrm{P}(\boldsymbol{\psi}^* \in B) \to 1$. Since $\boldsymbol{C}_{a(b)}$ and $\boldsymbol{D}_{a(b)}$ are continuous, for $\boldsymbol{\psi}^* \in B$, there exist C^* and D^* such that $|g_{ac}| \leq \sum_{b=1}^{q}|\widehat{\psi}_b - \psi_b|(N^{-1}C^*\boldsymbol{y}^\top\boldsymbol{y} + N^{-1}D^*)$. Note that $|\widehat{\psi}_b - \psi_b| = o_p(1)$ from Theorem 2.3 and $N^{-1}C^*\boldsymbol{y}^\top\boldsymbol{y} + N^{-1}D^* = O_p(1)$ from (C2). Thus, $|g_{ac}| = o_p(1)$, so that $\mathbf{col}_a(N^{-1}\ell_a) = -[O_p(1) + o_p(1)](\widehat{\boldsymbol{\psi}} - \boldsymbol{\psi})$, which means that $\widehat{\boldsymbol{\psi}} - \boldsymbol{\psi} = O_p(N^{-1/2})$. □

Proof of Theorem 2.5 Since the expansion (2.24) is rewritten as $\boldsymbol{0} = \mathbf{col}_a(\ell_a) - \boldsymbol{A}(\widehat{\boldsymbol{\psi}} - \boldsymbol{\psi}) + \{\mathbf{mat}_{ab}(\ell^{\dagger}_{a(b)}) + \boldsymbol{A}\}(\widehat{\boldsymbol{\psi}} - \boldsymbol{\psi})$, we have

$$\widehat{\boldsymbol{\psi}} - \boldsymbol{\psi} = \boldsymbol{A}^{-1}\mathbf{col}_a(\ell_a) + \boldsymbol{A}^{-1}\{\mathbf{mat}_{ab}(\ell^{\dagger}_{a(b)}) + \boldsymbol{A}\}(\widehat{\boldsymbol{\psi}} - \boldsymbol{\psi}). \tag{2.36}$$

It is noted that $\mathrm{E}[\ell_a\ell_b] = (\boldsymbol{B})_{ab}$ and that $\mathbf{mat}_{ab}(\ell_{a(b)}) + \boldsymbol{A} = O_p(N^{1/2})$ and $\widehat{\boldsymbol{\psi}} - \boldsymbol{\psi} = O_p(N^{-1/2})$ under conditions (C1) and (C2). Then it can be seen that $\mathbf{Cov}\,(\widehat{\boldsymbol{\psi}}) = \boldsymbol{A}^{-1}\mathbf{mat}_{ab}(\mathrm{E}[\ell_a\ell_b])\boldsymbol{A}^{-1} + o(N^{-1})$ under condition (C3). This leads to the expression in (2.30).

For part (2), from (2.36), it is noted that $\widehat{\boldsymbol{\psi}} - \boldsymbol{\psi} - \boldsymbol{A}^{-1}\mathbf{col}_a(\ell_a) = O_p(N^{-1})$. Also, note that $\mathrm{E}[\ell_{a(bc)}] = (\boldsymbol{E}_a)_{bc}$. The approximation in (2.36) is decomposed as

$$\begin{aligned}\widehat{\boldsymbol{\psi}} - \boldsymbol{\psi} =& \boldsymbol{A}^{-1}\mathbf{col}_a(\ell_a) + \boldsymbol{A}^{-1}\{\mathbf{mat}_{ab}(\ell_{a(b)}) + \boldsymbol{A}\}\boldsymbol{A}^{-1}\mathbf{col}_c(\ell_c)\\ &+ \frac{1}{2}\boldsymbol{A}^{-1}\mathbf{col}_a\{\mathbf{col}_b(\ell_b)^\top\boldsymbol{A}^{-1}\boldsymbol{E}_a\boldsymbol{A}^{-1}\mathbf{col}_c(\ell_c)\} + \boldsymbol{r}(\boldsymbol{y}),\end{aligned} \tag{2.37}$$

where $\boldsymbol{r}(\boldsymbol{y})$ is given in (2.29). Since $\mathrm{E}[\ell_a] = 0$ and $\mathrm{E}[\boldsymbol{r}(\boldsymbol{y})] = o(N^{-1})$, from condition (C4), it follows that

$$\begin{aligned}\mathrm{E}[\widehat{\boldsymbol{\psi}} - \boldsymbol{\psi}] =& \boldsymbol{A}^{-1}\mathrm{E}[\mathbf{mat}_{ab}(\ell_{a(b)})\boldsymbol{A}^{-1}\mathbf{col}_c(\ell_c)]\\ &+ \frac{1}{2}\boldsymbol{A}^{-1}\mathbf{col}_a\{\mathrm{E}[\mathbf{col}_b(\ell_b)^\top\boldsymbol{A}^{-1}\boldsymbol{E}_a\boldsymbol{A}^{-1}\mathbf{col}_c(\ell_c)]\} + o(N^{-1}).\end{aligned}$$

Using Lemma 2.1, we can see that

$$\begin{aligned}Ex[\ell_{a(b)}A^{bc}\ell_c] =& A^{bc}\mathrm{E}[\{\boldsymbol{y}^\top \boldsymbol{C}_{a(b)}\boldsymbol{y} - \mathrm{tr}\,(D_{a(b)})\}\{\boldsymbol{y}^\top \boldsymbol{C}_c \boldsymbol{y} - \mathrm{tr}\,(D_c)\}]\\ =& A^{bc}\{2\mathrm{tr}\,(C_{a(b)}\boldsymbol{\Sigma}\boldsymbol{C}_c\boldsymbol{\Sigma}) + \kappa(C_{a(b)}, \boldsymbol{C}_c)\},\end{aligned}$$

where A^{bc} is the (b, c)-element of the inverse matrix $\boldsymbol{A}^{-1}$. This gives

$$\begin{aligned}&\mathrm{E}[\mathbf{mat}_{ab}(\ell_{a(b)})\boldsymbol{A}^{-1}\mathbf{col}_c(\ell_c)]\\ =&\mathbf{col}_a\{2\mathrm{tr}\,[\boldsymbol{A}^{-1}\mathbf{mat}_{bc}(\boldsymbol{C}_{a(b)}\boldsymbol{\Sigma}\boldsymbol{C}_c\boldsymbol{\Sigma})] + \mathrm{tr}\,[\boldsymbol{A}^{-1}\mathbf{mat}_{bc}(\kappa(\boldsymbol{C}_{a(b)}, \boldsymbol{C}_c))]\}\\ =&\mathbf{col}_a\{\mathrm{tr}\,(\boldsymbol{F}_a\boldsymbol{A}^{-1})\}.\end{aligned}$$

Also, $\mathrm{E}[\ell_b A^{bc}(\boldsymbol{E}_a)_{cd}A^{de}\ell_e] = A^{bc}(\boldsymbol{E}_a)_{cd}A^{de}\{2\mathrm{tr}\,(C_b\boldsymbol{\Sigma}\boldsymbol{C}_e\boldsymbol{\Sigma}) + \kappa(C_b, \boldsymbol{C}_e)\}$, which gives

$$\mathrm{E}[\mathbf{col}_b(\ell_b)^\top \boldsymbol{A}^{-1}\boldsymbol{E}_a\boldsymbol{A}^{-1}\mathbf{col}_c(\ell_c)] = \mathrm{tr}\,(\boldsymbol{A}^{-1}\boldsymbol{E}_a\boldsymbol{A}^{-1}\boldsymbol{B}).$$

These provide the expression in (2.31) in Theorem 2.5. □

Proof of Theorem 2.6 Case of $\boldsymbol{W}_a = \boldsymbol{\Sigma}^{-1}\boldsymbol{\Sigma}_{(a)}\boldsymbol{\Sigma}^{-1}$. Note that

$$\boldsymbol{W}_{a(b)} = -\boldsymbol{\Sigma}^{-1}\boldsymbol{\Sigma}_{(b)}\boldsymbol{\Sigma}^{-1}\boldsymbol{\Sigma}_{(a)}\boldsymbol{\Sigma}^{-1} - \boldsymbol{\Sigma}^{-1}\boldsymbol{\Sigma}_{(a)}\boldsymbol{\Sigma}^{-1}\boldsymbol{\Sigma}_{(b)}\boldsymbol{\Sigma}^{-1} + \boldsymbol{\Sigma}^{-1}\boldsymbol{\Sigma}_{(ab)}\boldsymbol{\Sigma}^{-1}.$$

Then, $\mathrm{tr}\,(\boldsymbol{W}_a\boldsymbol{\Sigma}_{(b)})=\mathrm{tr}\,(\boldsymbol{\Sigma}^{-1}\boldsymbol{\Sigma}_{(a)}\boldsymbol{\Sigma}^{-1}\boldsymbol{\Sigma}_{(b)}) = (\boldsymbol{A})_{ab}$ and $(\boldsymbol{B})_{ab} = 2\mathrm{tr}\,(\boldsymbol{W}_a\boldsymbol{\Sigma}\boldsymbol{W}_b\boldsymbol{\Sigma})= 2\mathrm{tr}\,(\boldsymbol{\Sigma}^{-1}\boldsymbol{\Sigma}_{(a)}\boldsymbol{\Sigma}^{-1}\boldsymbol{\Sigma}_{(b)}) = 2(\boldsymbol{A})_{ab}$, which implies that $\boldsymbol{A}^{-1}\boldsymbol{B}\boldsymbol{A}^{-1} = 2\boldsymbol{A}^{-1}$. Thus, the covariance matrix of $\widehat{\boldsymbol{\psi}}$ is $2\boldsymbol{A}^{-1} + o(N^{-1})$. Moreover, note that

$$\begin{aligned}\mathrm{tr}\,(\boldsymbol{W}_{a(b)}\boldsymbol{\Sigma}\boldsymbol{W}_c\boldsymbol{\Sigma}) =& -2\mathrm{tr}\,(\boldsymbol{\Sigma}^{-1}\boldsymbol{\Sigma}_{(a)}\boldsymbol{\Sigma}^{-1}\boldsymbol{\Sigma}_{(b)}\boldsymbol{\Sigma}^{-1}\boldsymbol{\Sigma}_{(c)}) + \mathrm{tr}\,(\boldsymbol{\Sigma}^{-1}\boldsymbol{\Sigma}_{(ab)}\boldsymbol{\Sigma}^{-1}\boldsymbol{\Sigma}_{(c)}),\\ \mathrm{tr}\,(\boldsymbol{W}_{a(b)}\boldsymbol{\Sigma}_{(c)}) =& -2\mathrm{tr}\,(\boldsymbol{\Sigma}^{-1}\boldsymbol{\Sigma}_{(a)}\boldsymbol{\Sigma}^{-1}\boldsymbol{\Sigma}_{(b)}\boldsymbol{\Sigma}^{-1}\boldsymbol{\Sigma}_{(c)}) + \mathrm{tr}\,(\boldsymbol{\Sigma}^{-1}\boldsymbol{\Sigma}_{(ab)}\boldsymbol{\Sigma}^{-1}\boldsymbol{\Sigma}_{(c)}),\end{aligned}$$

which shows that $\boldsymbol{W}_a^{\mathrm{REML}}$ satisfies (2.33).

Case of $\boldsymbol{W}_a = (\boldsymbol{\Sigma}^{-1}\boldsymbol{\Sigma}_{(a)} + \boldsymbol{\Sigma}_{(a)}\boldsymbol{\Sigma}^{-1})/2$. From (2.32), it follows that $(\boldsymbol{A})_{ab} = \mathrm{tr}\,(\boldsymbol{\Sigma}^{-1}\boldsymbol{\Sigma}_{(a)}\boldsymbol{\Sigma}_{(b)})$ and $(\boldsymbol{B})_{ab} = \mathrm{tr}\,(\boldsymbol{\Sigma}_{(a)}\boldsymbol{\Sigma}_{(b)}) + \mathrm{tr}\,(\boldsymbol{\Sigma}^{-1}\boldsymbol{\Sigma}_{(a)}\boldsymbol{\Sigma}\boldsymbol{\Sigma}_{(b)})$. The asymptotic covariance matrix of $\widehat{\boldsymbol{\psi}}$ is $\boldsymbol{A}^{-1}\boldsymbol{B}\boldsymbol{A}^{-1}$, and the bias is derived from (2.31).

Case of $\boldsymbol{W}_a = \boldsymbol{\Sigma}_{(a)}$. Straightforward calculation shows that $(\boldsymbol{A})_{ab} = \mathrm{tr}\,(\boldsymbol{\Sigma}_{(a)}\boldsymbol{\Sigma}_{(b)})$ and $(\boldsymbol{B})_{ab} = 2\mathrm{tr}\,(\boldsymbol{\Sigma}_{(a)}\boldsymbol{\Sigma}\boldsymbol{\Sigma}_{(b)}\boldsymbol{\Sigma})$. The asymptotic covariance matrix of $\widehat{\boldsymbol{\psi}}$ is $\boldsymbol{A}^{-1}\boldsymbol{B}\boldsymbol{A}^{-1} + o(N^{-1})$. Moreover, since $W_{a(b)} = 0$, condition (2.33) holds. □

References

Battese GE, Harter RM, Fuller WA (1988) An error-components model for prediction of county crop areas using survey and satellite data. J. Am. Statist. Assoc. 83:28–36

Datta GS, Lahiri P (2000) A unified measure of uncertainty of estimated best linear unbiased predictors in small area estimation problems. Statist. Sinica 10:613–627

Fay RE, Herriot R (1979) Estimates of income for small places: An application of James-Stein procedures to census data. J. Am. Statist. Assoc. 74:269–277

Henderson CR (1950) Estimation of genetic parameters. Ann. Math. Statist. 21:309–310

Kubokawa T (2011) On measuring uncertainty of small area estimators with higher order accuracy. J. Japan Statist. Soc. 41:93–119
Kubokawa T, Sugasawa S, Tamae H, Chaudhuri S (2021) General unbiased estimating equations for variance components in linear mixed models. Japanese J. Statist. Data Sci. 4:841–859
McCulloch CE, Searle SR (2001) Generalized, Linear and Mixed Models. Wiley, New York
Prasad NGN, Rao JNK (1990) The estimation of the mean squared error of small area estimators. J. Am. Statist. Assoc. 85:163–171
Rao JNK, Yu M (1994) Small area estimation by combining time series and cross-sectional data. Candian J. Statist. 22:511–528
Rao CR, Kleffe J (1988) Estimation of variance components and applications. North-Holland, Amsterdam
Searle SR, Casella G, McCulloch CE (1992) Variance components. Wiley, New York
van der Vaart AW (1998) Asymptotic statistics. Cambridge University Press, Cambridge

Chapter 3
Measuring Uncertainty of Predictors

An important aspect of small area estimation is the assessment of accuracy of the predictors. Under the frequentist approach, this will be complicated due to the additional fluctuation induced by estimating unknown parameters in models. We here focus on two methods that are widely adopted in this context: estimators of mean squared error and confidence intervals.

3.1 EBLUP and the Mean Squared Error

The empirical (or estimated) best linear unbiased predictor (EBLUP) of $\theta = \boldsymbol{c}_1^\top \boldsymbol{\beta} + \boldsymbol{c}_2^\top \boldsymbol{v}$ is provided by substituting estimator $\widehat{\boldsymbol{\psi}}$ into the BLUP $\widetilde{\theta}^{\mathrm{BLUP}}(\boldsymbol{\psi})$ in (2.11), namely, the EBLUP is described as

$$\widehat{\theta}^{\mathrm{EBLUP}} = \widetilde{\theta}^{\mathrm{BLUP}}(\widehat{\boldsymbol{\psi}}) = \boldsymbol{c}_1^\top \widehat{\boldsymbol{\beta}} + \boldsymbol{c}_2^\top \widehat{\boldsymbol{R}}_v \boldsymbol{Z}^\top \widehat{\boldsymbol{\Sigma}}^{-1}(\boldsymbol{y} - \boldsymbol{X}\widehat{\boldsymbol{\beta}}), \tag{3.1}$$

where $\widehat{\boldsymbol{\beta}} = \widetilde{\boldsymbol{\beta}}(\widehat{\boldsymbol{\psi}})$, $\widehat{\boldsymbol{R}}_v = \boldsymbol{R}_v(\widehat{\boldsymbol{\psi}})$, $\widehat{\boldsymbol{R}}_e = \boldsymbol{R}_e(\widehat{\boldsymbol{\psi}})$, and $\widehat{\boldsymbol{\Sigma}} = \boldsymbol{\Sigma}(\widehat{\boldsymbol{\psi}})$. For notational simplicity, let

$$\begin{aligned}
\boldsymbol{K} &= \boldsymbol{R}_v \boldsymbol{Z}^\top \boldsymbol{\Sigma}^{-1}, \quad \boldsymbol{u} = \boldsymbol{y} - \boldsymbol{X}\boldsymbol{\beta}, \\
\boldsymbol{H} &= (\boldsymbol{X}^\top \boldsymbol{\Sigma}^{-1} \boldsymbol{X})^{-1} \boldsymbol{X}^\top \boldsymbol{\Sigma}^{-1}, \\
\boldsymbol{d} &= \boldsymbol{H}^\top \boldsymbol{c}_1 + (\boldsymbol{I} - \boldsymbol{H}^\top \boldsymbol{X}^\top) \boldsymbol{K}^\top \boldsymbol{c}_2,
\end{aligned} \tag{3.2}$$

and $\widehat{\boldsymbol{K}} = \widehat{\boldsymbol{R}}_v \boldsymbol{Z}^\top \widehat{\boldsymbol{\Sigma}}^{-1}$, $\widehat{\boldsymbol{H}} = (\boldsymbol{X}^\top \widehat{\boldsymbol{\Sigma}}^{-1} \boldsymbol{X})^{-1} \boldsymbol{X}^\top \widehat{\boldsymbol{\Sigma}}^{-1}$, and $\widehat{\boldsymbol{d}} = \widehat{\boldsymbol{H}}^\top \boldsymbol{c}_1 + (\boldsymbol{I} - \widehat{\boldsymbol{H}}^\top \boldsymbol{X}^\top) \widehat{\boldsymbol{K}}^\top \boldsymbol{c}_2$. Then, $\widetilde{\theta}^{\mathrm{BLUP}}(\boldsymbol{\psi}) = \boldsymbol{d}^\top \boldsymbol{u} + \boldsymbol{c}_1^\top \boldsymbol{\beta}$ and $\widehat{\theta}^{\mathrm{EBLUP}} = \widehat{\boldsymbol{d}}^\top \boldsymbol{u} + \boldsymbol{c}_1^\top \boldsymbol{\beta}$.

S. Sugasawa and T. Kubokawa, *Mixed-Effects Models and Small Area Estimation*, JSS Research Series in Statistics, https://doi.org/10.1007/978-981-19-9486-9_3

To measure uncertainty of EBLUP, we first evaluate the mean squared error (MSE). The MSE of $\widehat{\theta}^{\mathrm{EBLUP}}$ is decomposed as

$$\mathrm{MSE}(\boldsymbol{\psi}, \widehat{\theta}^{\mathrm{EBLUP}}) = G(\boldsymbol{\psi}) + G_{\mathrm{E}}(\boldsymbol{\psi}) + 2G_{\mathrm{C}}(\boldsymbol{\psi}), \tag{3.3}$$

where

$$\begin{aligned} G(\boldsymbol{\psi}) &= \mathrm{E}[\{\widetilde{\theta}^{\mathrm{BLUP}}(\boldsymbol{\psi}) - \theta\}^2], \\ G_{\mathrm{E}}(\boldsymbol{\psi}) &= \mathrm{E}[\{\widehat{\theta}^{\mathrm{BLUP}}(\widehat{\boldsymbol{\psi}}) - \widetilde{\theta}^{\mathrm{BLUP}}(\boldsymbol{\psi})\}^2], \\ G_{\mathrm{C}}(\boldsymbol{\psi}) &= \mathrm{E}[\{\widetilde{\theta}^{\mathrm{BLUP}}(\boldsymbol{\psi}) - \theta\}\{\widehat{\theta}^{\mathrm{EBLUP}} - \widetilde{\theta}^{\mathrm{BLUP}}(\boldsymbol{\psi})\}]. \end{aligned}$$

It is noted that $G(\boldsymbol{\psi})$ is the MSE of BLUP and $G_{\mathrm{E}}(\boldsymbol{\psi})$ measures the estimation error caused by estimating $\boldsymbol{\psi}$ with $\widehat{\boldsymbol{\psi}}$. Since $\theta = \boldsymbol{c}_1^\top \boldsymbol{\beta} + \boldsymbol{c}_2^\top \boldsymbol{v}$, we can calculate $G_1(\boldsymbol{\psi})$ as

$$G(\boldsymbol{\psi}) = \mathrm{E}[\{\boldsymbol{d}^\top(\boldsymbol{Z}\boldsymbol{v} + \boldsymbol{\epsilon}) - \boldsymbol{c}_2^\top \boldsymbol{v}\}^2] = \boldsymbol{d}^\top \boldsymbol{\Sigma} \boldsymbol{d} + \boldsymbol{c}_2^\top \boldsymbol{R}_v \boldsymbol{c}_2 - 2\boldsymbol{d}^\top \boldsymbol{Z}\boldsymbol{R}_v \boldsymbol{c}_2.$$

Noting that $\boldsymbol{H}\boldsymbol{\Sigma}\boldsymbol{H}^\top = (\boldsymbol{X}^\top \boldsymbol{\Sigma}^{-1}\boldsymbol{X})^{-1}$ and $\boldsymbol{H}\boldsymbol{\Sigma}(\boldsymbol{I} - \boldsymbol{H}^\top \boldsymbol{X}^\top) = \boldsymbol{0}$, we can express the term as

$$G(\boldsymbol{\psi}) = g_1(\boldsymbol{\psi}) + g_2(\boldsymbol{\psi}), \tag{3.4}$$

where

$$\begin{aligned} g_1(\boldsymbol{\psi}) &= \boldsymbol{c}_2^\top(\boldsymbol{R}_v - \boldsymbol{R}_v \boldsymbol{Z}^\top \boldsymbol{\Sigma}^{-1} \boldsymbol{Z}\boldsymbol{R}_v)\boldsymbol{c}_2, \\ g_2(\boldsymbol{\psi}) &= (\boldsymbol{c}_1^\top - \boldsymbol{c}_2^\top \boldsymbol{R}_v \boldsymbol{Z}^\top \boldsymbol{\Sigma}^{-1}\boldsymbol{X})(\boldsymbol{X}^\top \boldsymbol{\Sigma}^{-1}\boldsymbol{X})^{-1}(\boldsymbol{c}_1 - \boldsymbol{X}^\top \boldsymbol{\Sigma}^{-1}\boldsymbol{Z}\boldsymbol{R}_v \boldsymbol{c}_2). \end{aligned} \tag{3.5}$$

Using the equality $\boldsymbol{\Sigma}^{-1} = \boldsymbol{R}_e^{-1} - \boldsymbol{R}_e^{-1}\boldsymbol{Z}(\boldsymbol{R}_v^{-1} + \boldsymbol{Z}^\top \boldsymbol{R}_e^{-1}\boldsymbol{Z})^{-1}\boldsymbol{Z}^\top \boldsymbol{R}_e^{-1}$, we can rewrite $g_1(\boldsymbol{\psi})$ as

$$g_1(\boldsymbol{\psi}) = \boldsymbol{c}_2^\top(\boldsymbol{R}_v^{-1} + \boldsymbol{Z}^\top \boldsymbol{R}_e^{-1}\boldsymbol{Z})^{-1}\boldsymbol{c}_2. \tag{3.6}$$

Asymptotic approximations of $G_{\mathrm{E}}(\boldsymbol{\psi})$ and $G_{\mathrm{C}}(\boldsymbol{\psi})$ are studied in the next section.

3.2 Approximation of the MSE

We now evaluate $G_{\mathrm{E}}(\boldsymbol{\psi})$ and $G_{\mathrm{C}}(\boldsymbol{\psi})$ asymptotically. To this end, it is noted that

$$\begin{aligned} \widehat{\theta}^{\mathrm{EBLUP}} - \widetilde{\theta}^{\mathrm{BLUP}}(\boldsymbol{\psi}) &= (\widehat{\boldsymbol{d}} - \boldsymbol{d})^\top \boldsymbol{u} \\ &= \sum_{a=1}^{q} \boldsymbol{d}_{(a)}^\top \boldsymbol{u}(\widehat{\psi}_a - \psi_a) + \left\{(\widehat{\boldsymbol{d}} - \boldsymbol{d})^\top \boldsymbol{u} - \sum_{a=1}^{q} \boldsymbol{d}_{(a)}^\top \boldsymbol{u}(\widehat{\psi}_a - \psi_a)\right\}, \end{aligned} \tag{3.7}$$

where $\boldsymbol{d}_{(a)}$ is written as $\boldsymbol{d}_{(a)}^\top \boldsymbol{u} = \boldsymbol{c}_2^\top \boldsymbol{K}_{(a)}\boldsymbol{u} + \boldsymbol{m}_a^\top \boldsymbol{u}$ for

$$\boldsymbol{m}_a^\top = (\boldsymbol{c}_1^\top - \boldsymbol{c}_2^\top \boldsymbol{K}\boldsymbol{X})\boldsymbol{H}_{(a)} - \boldsymbol{c}_2^\top \boldsymbol{K}_{(a)}\boldsymbol{X}\boldsymbol{H}.$$

It is noted that $\boldsymbol{c}_2^\top \boldsymbol{K}_{(a)}\boldsymbol{u} = O_p(1)$ and $\boldsymbol{m}_a^\top \boldsymbol{u} = O_p(N^{-1/2})$ from condition (B1) given below. Let $\mathrm{E}[\cdot \mid \boldsymbol{c}_2^\top \boldsymbol{K}_{(a)}\boldsymbol{u}, \boldsymbol{c}_2^\top \boldsymbol{K}_{(b)}\boldsymbol{u}]$ be the conditional expectation given $\boldsymbol{c}_2^\top \boldsymbol{K}_{(a)}\boldsymbol{u}$ and $\boldsymbol{c}_2^\top \boldsymbol{K}_{(b)}\boldsymbol{u}$. Define $R_0(\boldsymbol{y})$ by

$$R_0(\boldsymbol{y}) = \mathrm{E}[(\widehat{\psi}_a - \psi_a)(\widehat{\psi}_b - \psi_b) \mid \boldsymbol{c}_2^\top \boldsymbol{K}_{(a)}\boldsymbol{u}, \boldsymbol{c}_2^\top \boldsymbol{K}_{(b)}\boldsymbol{u}] - \mathrm{E}[(\widehat{\psi}_a - \psi_a)(\widehat{\psi}_b - \psi_b)].$$

Assume the conditions.

(B1) $\boldsymbol{c}_2^\top \boldsymbol{K}_{(a)}\boldsymbol{K}_{(a)}^\top \boldsymbol{c}_2 = O(1)$, $\boldsymbol{m}_a^\top \boldsymbol{m}_a = O(N^{-1})$ and $\mathrm{E}[(\boldsymbol{m}_a^\top \boldsymbol{u})^2(\widehat{\psi}_a - \psi_a)^2] = o(N^{-1})$.

(B2) $\mathrm{Cov}(\widehat{\psi}_a, \widehat{\psi}_b) = O(N^{-1})$, $\mathrm{E}[\{(\widehat{\boldsymbol{d}} - \boldsymbol{d})^\top \boldsymbol{u} - \sum_{a=1}^q \boldsymbol{d}_{(a)}^\top \boldsymbol{u}(\widehat{\psi}_a - \psi_a)\}^2] = o(N^{-1})$, and $\mathrm{E}[\boldsymbol{c}_2^\top \boldsymbol{K}_{(a)}\boldsymbol{u} \cdot \boldsymbol{c}_2^\top \boldsymbol{K}_{(b)}\boldsymbol{u} R_0(\boldsymbol{y})] = o(N^{-1})$.

(B3) $\mathrm{E}[\{\widehat{\boldsymbol{d}} - \boldsymbol{d} - \sum_a \boldsymbol{d}_{(a)}(\widehat{\psi}_a - \psi_a)\}^\top \boldsymbol{u}(\boldsymbol{d}^\top \boldsymbol{u} - \boldsymbol{c}_2^\top \boldsymbol{v})] = o(N^{-1})$.

Theorem 3.1 *Under conditions* (B1)–(B3), *it holds that* $G_{\mathrm{E}}(\boldsymbol{\psi}) = g_3(\boldsymbol{\psi}) + o(N^{-1})$ *and* $G_{\mathrm{C}}(\boldsymbol{\psi}) = g_4(\boldsymbol{\psi}) + o(N^{-1})$, *where*

$$g_3(\boldsymbol{\psi}) = \sum_{a,b} \boldsymbol{d}_{(a)}^\top \boldsymbol{\Sigma} \boldsymbol{d}_{(b)} \mathrm{Cov}(\widehat{\psi}_a, \widehat{\psi}_b), \tag{3.8}$$

$$g_4(\boldsymbol{\psi}) = \sum_a \mathrm{E}[\boldsymbol{d}_{(a)}^\top \boldsymbol{u}(\widehat{\psi}_a - \psi_a)(\boldsymbol{d}^\top \boldsymbol{u} - \boldsymbol{c}_2^\top \boldsymbol{v})]. \tag{3.9}$$

Thus, the MSE of $\widehat{\theta}^{\mathrm{EBLUP}}$ *is approximated as*

$$\mathrm{MSE}(\boldsymbol{\psi}, \widehat{\theta}^{\mathrm{EBLUP}}) = g_1(\boldsymbol{\psi}) + g_2(\boldsymbol{\psi}) + g_3(\boldsymbol{\psi}) + 2g_4(\boldsymbol{\psi}) + o(N^{-1}), \tag{3.10}$$

where $g_1(\boldsymbol{\psi})$ *and* $g_2(\boldsymbol{\psi})$ *are given in (3.5).*

It is noted that $g_1(\boldsymbol{\psi}) = O(1)$, while $g_i(\boldsymbol{\psi}) = O(N^{-1})$ for $i = 2, 3, 4$. Also, $g_1(\boldsymbol{\psi})$ and $g_2(\boldsymbol{\psi})$ do not depend on estimator $\widehat{\boldsymbol{\psi}}$, while $g_3(\boldsymbol{\psi})$ and $g_4(\boldsymbol{\psi})$ depend on $\widehat{\boldsymbol{\psi}}$.

For the estimator $\widehat{\boldsymbol{\psi}}$ defined in (2.20) or (2.22), we can obtain the expressions of $g_3(\boldsymbol{\psi})$ and $g_4(\boldsymbol{\psi})$. Let

$$\begin{aligned} R_1(\boldsymbol{u}, \boldsymbol{v}) =& \mathrm{E}[\{\ell_{b(c)} + (\boldsymbol{A})_{bc}\}\ell_d \mid \boldsymbol{c}_2^\top \boldsymbol{K}_{(a)}\boldsymbol{u}, \boldsymbol{c}_2^\top \boldsymbol{K}\boldsymbol{u}, \boldsymbol{c}_2^\top \boldsymbol{v}] - \mathrm{E}[\ell_{b(c)}\ell_d], \\ R_2(\boldsymbol{u}, \boldsymbol{v}) =& \mathrm{E}[\ell_c\ell_d \mid \boldsymbol{c}_2^\top \boldsymbol{K}_{(a)}\boldsymbol{u}, \boldsymbol{c}_2^\top \boldsymbol{K}\boldsymbol{u}, \boldsymbol{c}_2^\top \boldsymbol{v}] - \mathrm{E}[\ell_c\ell_d], \end{aligned} \tag{3.11}$$

for $a, b, c, d \in \{1, \ldots, q\}$. Assume the following condition.

(B4) $\mathrm{E}[\boldsymbol{c}_2^\top \boldsymbol{K}_{(a)}\boldsymbol{u} \cdot (\boldsymbol{r}(\boldsymbol{y}))_a(\boldsymbol{d}^\top \boldsymbol{u} - \boldsymbol{c}_2^\top \boldsymbol{v})] = o(N^{-1})$ for $\boldsymbol{r}(\boldsymbol{y})$ given in (2.29). $R_1(\boldsymbol{u}, \boldsymbol{v})$ and $R_2(\boldsymbol{u}, \boldsymbol{v})$ satisfy that $R_1(\boldsymbol{u}, \boldsymbol{v}) = o_p(N)$ and $R_2(\boldsymbol{u}, \boldsymbol{v}) = o_p(N)$. Also, $\mathrm{E}[R_1(\boldsymbol{u}, \boldsymbol{v})\boldsymbol{c}_2^\top \boldsymbol{K}_{(a)}\boldsymbol{u} \cdot (\boldsymbol{d}^\top \boldsymbol{u} - \boldsymbol{c}_2^\top \boldsymbol{v})] = o(N)$ and $\mathrm{E}[R_2(\boldsymbol{u}, \boldsymbol{v})\boldsymbol{c}_2^\top \boldsymbol{K}_{(a)}\boldsymbol{u} \cdot (\boldsymbol{d}^\top \boldsymbol{u} - \boldsymbol{c}_2^\top \boldsymbol{v})] = o(N)$ are satisfied.

Theorem 3.2 *Assume that conditions* (C1)–(C5) *and* (B1)–(B4) *hold. For the estimator* $\widehat{\boldsymbol{\psi}}$ *defined in (2.20) or (2.22), the functions* $g_3(\boldsymbol{\psi})$ *and* $g_4(\boldsymbol{\psi})$ *are approximated as*

$$g_3(\boldsymbol{\psi}) = \sum_{a,b}^{q} \boldsymbol{c}_2^\top \boldsymbol{K}_{(a)} \boldsymbol{\Sigma} \boldsymbol{K}_{(a)}^\top \boldsymbol{c}_2 (\boldsymbol{A}^{-1}\boldsymbol{B}\boldsymbol{A}^{-1})_{ab} + o(N^{-1}), \tag{3.12}$$

$$g_4(\boldsymbol{\psi}) = \sum_{a,b}^{q} (\boldsymbol{A})^{ab} \Big\{ \kappa_e \sum_{j=1}^{N} (\boldsymbol{R}_e^{1/2} \boldsymbol{W}_b \boldsymbol{R}_e^{1/2})_{jj} \cdot (\boldsymbol{R}_e^{1/2} \boldsymbol{K}_{(a)}^\top \boldsymbol{c}_2 \boldsymbol{c}_2^\top \boldsymbol{K} \boldsymbol{R}_e^{1/2})_{jj} \tag{3.13}$$
$$+ \kappa_v \sum_{j=1}^{m} (\boldsymbol{R}_v^{1/2} \boldsymbol{Z}^\top \boldsymbol{W}_b \boldsymbol{Z} \boldsymbol{R}_v^{1/2})_{jj} \cdot (\boldsymbol{R}_v^{1/2} \boldsymbol{Z}^\top \boldsymbol{K}_{(a)}^\top \boldsymbol{c}_2 \boldsymbol{c}_2^\top (\boldsymbol{K}\boldsymbol{Z} - \boldsymbol{I}_m) \boldsymbol{R}_v^{1/2})_{jj} \Big\},$$

where the (a, b) *elements of* $\boldsymbol{A}$ *and* $\boldsymbol{B}$ *are* $(\boldsymbol{A})_{ab} = \mathrm{tr}\,(\boldsymbol{W}_a \boldsymbol{\Sigma}_{(b)}) + O(1)$ *and* $(\boldsymbol{B})_{ab} = 2\mathrm{tr}\,(\boldsymbol{W}_a \boldsymbol{\Sigma} \boldsymbol{W}_b \boldsymbol{\Sigma}) + \kappa(\boldsymbol{W}_a, \boldsymbol{W}_b) + O(1)$. *Under the normality of* $\boldsymbol{v}$ *and* $\boldsymbol{\epsilon}$, *it holds that* $g_4(\boldsymbol{\psi}) = 0$.

We provide the proofs of the theorems below.

Proof of Theorem 3.1 From the expression in (3.7), the second term in (3.3) is

$$\mathrm{E}\big[\{\widehat{\theta}^{\mathrm{EBLUP}} - \widetilde{\theta}^{\mathrm{BLUP}}(\boldsymbol{\psi})\}^2\big] = \mathrm{E}\Big[\Big\{\sum_{a=1}^{q} \boldsymbol{d}_{(a)}^\top \boldsymbol{u}(\widehat{\psi}_a - \psi_a)\Big\}^2\Big]$$
$$+ \mathrm{E}\Big[\Big\{(\widehat{\boldsymbol{d}} - \boldsymbol{d})^\top \boldsymbol{u} - \sum_{a=1}^{q} \boldsymbol{d}_{(a)}^\top \boldsymbol{u}(\widehat{\psi}_a - \psi_a)\Big\}^2\Big]$$
$$+ 2\mathrm{E}\Big[\Big\{\sum_{a=1}^{q} \boldsymbol{d}_{(a)}^\top \boldsymbol{u}(\widehat{\psi}_a - \psi_a)\Big\}\Big\{(\widehat{\boldsymbol{d}} - \boldsymbol{d})^\top \boldsymbol{u} - \sum_{a=1}^{q} \boldsymbol{d}_{(a)}^\top \boldsymbol{u}(\widehat{\psi}_a - \psi_a)\Big\}\Big].$$

From condition (B2) and the Cauchy–Schwarz inequality, it follows that

$$\mathrm{E}[\{\widehat{\theta}^{\mathrm{EBLUP}} - \widetilde{\theta}^{\mathrm{BLUP}}(\boldsymbol{\psi})\}^2] = \mathrm{E}\Big[\Big\{\sum_{a=1}^{q} \boldsymbol{d}_{(a)}^\top \boldsymbol{u}(\widehat{\psi}_a - \psi_a)\Big\}^2\Big] + o(N^{-1}),$$

if $\mathrm{E}[\{\sum_{a=1}^{q} \boldsymbol{d}_{(a)}^\top \boldsymbol{u}(\widehat{\psi}_a - \psi_a)\}^2] = O(N^{-1})$. Since $\boldsymbol{d}_{(a)}^\top \boldsymbol{u} = \boldsymbol{c}_2^\top \boldsymbol{K}_{(a)} \boldsymbol{u} + \boldsymbol{m}_a^\top \boldsymbol{u}$, we have

$$E\Big[\Big\{\sum_{a=1}^{q} \boldsymbol{d}_{(a)}^\top \boldsymbol{u}(\widehat{\psi}_a - \psi_a)\Big\}^2\Big] = \sum_{a,b} \Big\{\mathrm{E}[\boldsymbol{c}_2^\top \boldsymbol{K}_{(a)} \boldsymbol{u} \cdot \boldsymbol{c}_2^\top \boldsymbol{K}_{(b)} \boldsymbol{u}(\widehat{\psi}_a - \psi_a)(\widehat{\psi}_b - \psi_b)]$$
$$+ \mathrm{E}[\boldsymbol{c}_2^\top \boldsymbol{K}_{(a)} \boldsymbol{u} \cdot \boldsymbol{m}_b^\top \boldsymbol{u}(\widehat{\psi}_a - \psi_a)(\widehat{\psi}_b - \psi_b)] + \mathrm{E}[\boldsymbol{m}_a^\top \boldsymbol{u} \cdot \boldsymbol{m}_b^\top \boldsymbol{u}(\widehat{\psi}_a - \psi_a)(\widehat{\psi}_b - \psi_b)]\Big\}.$$

From condition (B2),

$$
\begin{aligned}
&\mathrm{E}[\boldsymbol{c}_2^\top \boldsymbol{K}_{(a)}\boldsymbol{u}\cdot \boldsymbol{c}_2^\top \boldsymbol{K}_{(b)}\boldsymbol{u}(\widehat{\psi}_a-\psi_a)(\widehat{\psi}_b-\psi_b)]\\
=&\mathrm{E}[\boldsymbol{c}_2^\top \boldsymbol{K}_{(a)}\boldsymbol{u}\cdot \boldsymbol{c}_2^\top \boldsymbol{K}_{(b)}\boldsymbol{u}]\mathrm{E}[(\widehat{\psi}_a-\psi_a)(\widehat{\psi}_b-\psi_b)]+\mathrm{E}[\boldsymbol{c}_2^\top \boldsymbol{K}_{(a)}\boldsymbol{u}\cdot \boldsymbol{c}_2^\top \boldsymbol{K}_{(b)}\boldsymbol{u}R_0(\boldsymbol{y})]\\
=&\boldsymbol{c}_2^\top \boldsymbol{K}_{(a)}\boldsymbol{K}_{(b)}^\top \boldsymbol{c}_2\mathrm{Cov}(\widehat{\psi}_a,\widehat{\psi}_b)+o(N^{-1}),
\end{aligned}
$$

which is of order $O(N^{-1})$. From the Cauchy–Schwarz inequality and condition (B1), it follows that

$$
\begin{aligned}
&\mathrm{E}[\boldsymbol{m}_a^\top \boldsymbol{u}\cdot \boldsymbol{m}_b^\top \boldsymbol{u}(\widehat{\psi}_a-\psi_a)(\widehat{\psi}_b-\psi_b)]\\
&\le \big\{\mathrm{E}[(\boldsymbol{m}_a^\top \boldsymbol{u})^2(\widehat{\psi}_a-\psi_a)^2]\big\}^{1/2}\big\{\mathrm{E}[(\boldsymbol{m}_b^\top \boldsymbol{u})^2(\widehat{\psi}_b-\psi_b)^2]\big\}^{1/2},
\end{aligned}
$$

which is of order $o(N^{-1})$. Also from the Cauchy–Schwarz inequality,

$$
\begin{aligned}
&\mathrm{E}[\boldsymbol{c}_2^\top \boldsymbol{K}_{(a)}\boldsymbol{u}\cdot \boldsymbol{m}_b^\top \boldsymbol{u}(\widehat{\psi}_a-\psi_a)(\widehat{\psi}_b-\psi_b)]\\
\le&\{\mathrm{E}[(\boldsymbol{c}_2^\top \boldsymbol{K}_{(a)}\boldsymbol{u})^2(\widehat{\psi}_a-\psi_a)^2]\}^{1/2}\{\mathrm{E}[(\boldsymbol{m}_b^\top \boldsymbol{u})^2(\widehat{\psi}_b-\psi_b)^2]\}^{1/2},
\end{aligned}
$$

which is of order $o(N^{-1})$ from the above arguments. Thus,

$$
\mathrm{E}[\{\widehat{\theta}^{\mathrm{EBLUP}}-\widetilde{\theta}^{\mathrm{BLUP}}(\boldsymbol{\psi})\}^2]=g_3(\boldsymbol{\psi})+o(N^{-1}). \tag{3.14}
$$

From condition (B3), the third term in (3.3) is

$$
\begin{aligned}
&\mathrm{E}[\{\widehat{\theta}^{\mathrm{EBLUP}}-\widetilde{\theta}^{\mathrm{BLUP}}(\boldsymbol{\psi})\}\{\widetilde{\theta}^{\mathrm{BLUP}}(\boldsymbol{\psi})-\theta\}]\\
=&\mathrm{E}[(\widehat{\boldsymbol{d}}-\boldsymbol{d})^\top \boldsymbol{u}(\boldsymbol{d}^\top \boldsymbol{u}-\boldsymbol{c}_2^\top \boldsymbol{v})]\\
=&g_4(\boldsymbol{\psi})+\mathrm{E}\Big[\Big\{\widehat{\boldsymbol{d}}-\boldsymbol{d}-\sum_a \boldsymbol{d}_{(a)}(\widehat{\psi}_a-\psi_a)\Big\}^\top \boldsymbol{u}(\boldsymbol{d}^\top \boldsymbol{u}-\boldsymbol{c}_2^\top \boldsymbol{v})\Big]\\
=&g_4(\boldsymbol{\psi})+o(N^{-1}).
\end{aligned} \tag{3.15}
$$

Combining (3.3), (3.5), (3.14), and (3.15), we obtain the approximation of the MSE of EBLUP. □

***Proof of Theorem* 3.2** The expression of $g_3(\boldsymbol{\psi})$ in (3.12) is derived from Theorem 2.5. For $\boldsymbol{r}(\boldsymbol{y})$ given in (2.29), from (2.37), it is noted that

$$
\begin{aligned}
\widehat{\psi}_a-\psi_a=&\sum_{b=1}^q(\boldsymbol{A})^{ab}\ell_b+\sum_{b,c,d}(\boldsymbol{A})^{ab}\{\ell_{b(c)}+(\boldsymbol{A})_{bc}\}(\boldsymbol{A})^{cd}\ell_d\\
&+\frac{1}{2}\sum_{b=1}^q(\boldsymbol{A})^{ab}\mathbf{col}_b(\ell_c)^\top \boldsymbol{A}^{-1}\boldsymbol{E}_b\boldsymbol{A}^{-1}\mathbf{col}_c(\ell_c)+(\boldsymbol{r}(\boldsymbol{y}))_a,
\end{aligned}
$$

which implies that $g_4(\boldsymbol{\psi})=\sum_{a,b}I_{1ab}+\sum_{a,b,c,d}I_{2abcd}+2^{-1}\sum_{a,b}I_{3ab}+\sum_a I_{4a}$, where for $\boldsymbol{u}=\boldsymbol{y}-\boldsymbol{X}\boldsymbol{\beta}$,

$$
\begin{aligned}
I_{1ab} =& \mathrm{E}[\boldsymbol{c}_2^\top \boldsymbol{K}_{(a)}\boldsymbol{u} \cdot (\boldsymbol{A})^{ab}\ell_b(\boldsymbol{d}^\top \boldsymbol{u} - \boldsymbol{c}_2^\top \boldsymbol{v})],\\
I_{2abcd} =& \mathrm{E}[\boldsymbol{c}_2^\top \boldsymbol{K}_{(a)}\boldsymbol{u} \cdot (\boldsymbol{A})^{ab}\{\ell_{b(c)} + (\boldsymbol{A})_{bc}\}(\boldsymbol{A})^{cd}\ell_d(\boldsymbol{d}^\top \boldsymbol{u} - \boldsymbol{c}_2^\top \boldsymbol{v})],\\
I_{3ab} =& \mathrm{E}[\boldsymbol{c}_2^\top \boldsymbol{K}_{(a)}\boldsymbol{u} \cdot (\boldsymbol{A})^{ab}\mathbf{col}_c(\ell_c)^\top \boldsymbol{A}^{-1}\boldsymbol{E}_b\boldsymbol{A}^{-1}\mathbf{col}_c(\ell_c)(\boldsymbol{d}^\top \boldsymbol{u} - \boldsymbol{c}_2^\top \boldsymbol{v})],\\
I_{4a} =& \mathrm{E}[\boldsymbol{c}_2^\top \boldsymbol{K}_{(a)}\boldsymbol{u} \cdot (\boldsymbol{r}(\boldsymbol{y}))_a(\boldsymbol{d}^\top \boldsymbol{u} - \boldsymbol{c}_2^\top \boldsymbol{v})].
\end{aligned}
$$

Since $\ell_b = \boldsymbol{u}^\top \boldsymbol{C}_b\boldsymbol{u} - \mathrm{tr}\,\boldsymbol{D}_b$ for $\boldsymbol{D}_b = \boldsymbol{C}_b\boldsymbol{\Sigma}$, Lemma 2.1 is used to rewrite I_{1ab} as

$$
\begin{aligned}
&(\boldsymbol{A})^{ab}\Big\{\mathrm{E}[(\boldsymbol{u}^\top \boldsymbol{C}_b\boldsymbol{u} - \mathrm{tr}\,\boldsymbol{D}_b)\boldsymbol{u}^\top \boldsymbol{K}_{(a)}^\top \boldsymbol{c}_2\boldsymbol{d}^\top \boldsymbol{u}] - \mathrm{E}[\{\boldsymbol{u}^\top \boldsymbol{C}_b\boldsymbol{u} - \mathrm{tr}\,(\boldsymbol{D}_b)\}\boldsymbol{u}^\top \boldsymbol{K}_{(a)}^\top \boldsymbol{c}_2\boldsymbol{c}_2^\top \boldsymbol{v}]\Big\}\\
=&(\boldsymbol{A})^{ab}\Big\{\mathrm{E}[\boldsymbol{u}^\top \boldsymbol{C}_b\boldsymbol{u}\boldsymbol{u}^\top \boldsymbol{K}_{(a)}^\top \boldsymbol{c}_2\boldsymbol{d}^\top \boldsymbol{u}] - \mathrm{tr}\,(\boldsymbol{D}_b)\mathrm{E}[\boldsymbol{u}^\top \boldsymbol{K}_{(a)}^\top \boldsymbol{c}_2\boldsymbol{d}^\top \boldsymbol{u}]\\
&- \mathrm{E}[(\boldsymbol{v}^\top \boldsymbol{Z}^\top \boldsymbol{C}_b\boldsymbol{Z}\boldsymbol{v} + 2\boldsymbol{v}^\top \boldsymbol{Z}^\top \boldsymbol{C}_b\boldsymbol{\epsilon} + \boldsymbol{\epsilon}^\top \boldsymbol{C}_b\boldsymbol{\epsilon})(\boldsymbol{v}^\top \boldsymbol{Z}^\top \boldsymbol{K}_{(a)}^\top \boldsymbol{c}_2\boldsymbol{c}_2^\top \boldsymbol{v} + \boldsymbol{\epsilon}^\top \boldsymbol{K}_{(a)}^\top \boldsymbol{c}_2\boldsymbol{c}_2^\top \boldsymbol{v})]\Big\},
\end{aligned}
$$

which is further rewritten as

$$
\begin{aligned}
&(\boldsymbol{A})^{ab}\Big\{\{2\mathrm{tr}\,(\boldsymbol{C}_b\boldsymbol{\Sigma}\boldsymbol{K}_{(a)}^\top \boldsymbol{c}_2\boldsymbol{d}^\top \boldsymbol{\Sigma}) + \kappa(\boldsymbol{C}_b, \boldsymbol{K}_{(a)}^\top \boldsymbol{c}_2\boldsymbol{d}^\top)\}\\
&- \{2\boldsymbol{c}_2^\top \boldsymbol{R}_v\boldsymbol{Z}^\top \boldsymbol{C}_b\boldsymbol{Z}\boldsymbol{R}_v\boldsymbol{Z}^\top \boldsymbol{K}_{(a)}^\top \boldsymbol{c}_2 + \boldsymbol{c}_2^\top \boldsymbol{R}_v\boldsymbol{Z}^\top \boldsymbol{K}_{(a)}^\top \boldsymbol{c}_2\mathrm{tr}\,(\boldsymbol{Z}^\top \boldsymbol{C}_b\boldsymbol{Z}\boldsymbol{R}_v)\\
&+ h_v(\boldsymbol{Z}^\top \boldsymbol{C}_b\boldsymbol{Z}, \boldsymbol{Z}^\top \boldsymbol{K}_{(a)}^\top \boldsymbol{c}_2\boldsymbol{c}_2^\top) + \boldsymbol{c}_2^\top \boldsymbol{R}_v\boldsymbol{Z}^\top \boldsymbol{K}_{(a)}^\top \boldsymbol{c}_2\mathrm{tr}\,(\boldsymbol{C}_b\boldsymbol{R}_e)\\
&+ 2\boldsymbol{c}_2^\top \boldsymbol{R}_v\boldsymbol{Z}^\top \boldsymbol{C}_b\boldsymbol{R}_e\boldsymbol{R}_e\boldsymbol{K}_{(a)}^\top \boldsymbol{c}_2 - \boldsymbol{c}_2^\top \boldsymbol{R}_v\boldsymbol{Z}^\top \boldsymbol{K}_{(a)}^\top \boldsymbol{c}_2\mathrm{tr}\,(\boldsymbol{D}_b)\}\Big\},
\end{aligned}
$$

where $h_v(\boldsymbol{C}, \boldsymbol{D}) = \kappa_v \sum_{j=1}^m (\boldsymbol{R}_v^{1/2}\boldsymbol{C}\boldsymbol{R}_v^{1/2})_{jj} \cdot (\boldsymbol{R}_v^{1/2}\boldsymbol{D}\boldsymbol{R}_v^{1/2})_{jj}$ and $\kappa(\boldsymbol{C}, \boldsymbol{D})$ is given in (2.27). Noting that $\boldsymbol{\Sigma} = \boldsymbol{Z}\boldsymbol{R}_v\boldsymbol{Z}^\top + \boldsymbol{R}_e$, we can see that

$$
\begin{aligned}
I_{1ab} =&(\boldsymbol{A})^{ab}\{\kappa(\boldsymbol{C}_b, \boldsymbol{K}_{(a)}^\top \boldsymbol{c}_2\boldsymbol{d}^\top) - h_v(\boldsymbol{Z}^\top \boldsymbol{C}_b\boldsymbol{Z}, \boldsymbol{Z}^\top \boldsymbol{K}_{(a)}^\top \boldsymbol{c}_2\boldsymbol{c}_2^\top)\}\\
=&(\boldsymbol{A})^{ab}\{\kappa(\boldsymbol{C}_b, \boldsymbol{K}_{(a)}^\top \boldsymbol{c}_2\boldsymbol{c}_2^\top \boldsymbol{K}) - h_v(\boldsymbol{Z}^\top \boldsymbol{C}_b\boldsymbol{Z}, \boldsymbol{Z}^\top \boldsymbol{K}_{(a)}^\top \boldsymbol{c}_2\boldsymbol{c}_2^\top)\} + o(N^{-1}). \quad (3.16)
\end{aligned}
$$

Concerning the evaluation of I_{2abcd}, it is written as

$$
\begin{aligned}
I_{2abcd} =&(\boldsymbol{A})^{ab}(\boldsymbol{A})^{cd}\mathrm{E}[\ell_{b(c)}\ell_d]\mathrm{E}[\boldsymbol{u}^\top \boldsymbol{K}_{(a)}^\top \boldsymbol{c}_2(\boldsymbol{d}^\top \boldsymbol{u} - \boldsymbol{c}_2^\top \boldsymbol{v})]\\
&+ (\boldsymbol{A})^{ab}(\boldsymbol{A})^{cd}\mathrm{E}[R_1(\boldsymbol{u}, \boldsymbol{v})\boldsymbol{c}_2^\top \boldsymbol{K}_{(a)}\top \boldsymbol{u}(\boldsymbol{d}^\top \boldsymbol{u} - \boldsymbol{c}_2^\top \boldsymbol{v})].
\end{aligned}
$$

From (B4), the second term in RHS is of order $o(N^{-1})$. Note that

$$
\begin{aligned}
\mathrm{E}[\boldsymbol{c}_2^\top \boldsymbol{K}_{(a)}\boldsymbol{u}(\boldsymbol{d}^\top \boldsymbol{u} - \boldsymbol{c}_2^\top \boldsymbol{v})] =&\boldsymbol{c}_2^\top \boldsymbol{K}_{(a)}\boldsymbol{\Sigma}\boldsymbol{d} - \boldsymbol{c}_2^\top \boldsymbol{K}_{(a)}\boldsymbol{Z}\boldsymbol{R}_v\boldsymbol{c}_2\\
=&\boldsymbol{c}_2^\top \boldsymbol{K}_{(a)}\boldsymbol{\Sigma}\boldsymbol{H}^\top(\boldsymbol{c}_1 - \boldsymbol{X}^\top \boldsymbol{K}^\top \boldsymbol{c}_2),
\end{aligned}
$$

which is of order $O(N^{-1})$. Using Lemma 2.1, we can demonstrate that $\mathrm{E}[\ell_{b(c)}\ell_d] = 2\mathrm{tr}\,(\boldsymbol{C}_{b(c)}\boldsymbol{\Sigma}\boldsymbol{C}_d\boldsymbol{\Sigma}) + \kappa(\boldsymbol{C}_{b(c)}, \boldsymbol{C}_d)$, which yields

$$\begin{aligned} I_{2abcd} =&(\boldsymbol{A})^{ab}(\boldsymbol{A})^{cd}\{2\mathrm{tr}\,(\boldsymbol{C}_{b(c)}\boldsymbol{\Sigma}\boldsymbol{C}_d\boldsymbol{\Sigma}) + \kappa(\boldsymbol{C}_{b(c)}, \boldsymbol{C}_d)\}(\boldsymbol{c}_1^\top - \boldsymbol{c}_2^\top \boldsymbol{K}\boldsymbol{X})\boldsymbol{H}\boldsymbol{\Sigma}\boldsymbol{d}_{(a)} \\ &+ o(N^{-1}). \end{aligned}$$

Since $(\boldsymbol{c}_1^\top - \boldsymbol{c}_2^\top \boldsymbol{K}\boldsymbol{X})\boldsymbol{H}\boldsymbol{\Sigma}\boldsymbol{K}_{(a)}^\top \boldsymbol{c}_2 = O(N^{-1})$ and $(\boldsymbol{A})^{ab} = O(N^{-1})$, we have

$$I_{2abcd} = o(N^{-1}), \tag{3.17}$$

For I_{3ab}, it is written as

$$I_{3ab} = \sum_{c,d}(\boldsymbol{A})^{ab}(\boldsymbol{A}^{-1}\boldsymbol{E}_b\boldsymbol{A}^{-1})_{cd}\mathrm{E}[\ell_c\ell_d\boldsymbol{c}_2^\top \boldsymbol{K}_{(a)}\boldsymbol{u}(\boldsymbol{d}^\top \boldsymbol{u} - \boldsymbol{c}_2^\top \boldsymbol{v})].$$

From condition (B4), it follows that

$$\begin{aligned} &\mathrm{E}[\ell_c\ell_d\boldsymbol{c}_2^\top \boldsymbol{K}_{(a)}\boldsymbol{u}(\boldsymbol{d}^\top \boldsymbol{u} - \boldsymbol{c}_2^\top \boldsymbol{v})] \\ =&\mathrm{E}[\ell_c\ell_d]\mathrm{E}[\boldsymbol{c}_2^\top \boldsymbol{K}_{(a)}\boldsymbol{u}(\boldsymbol{d}^\top \boldsymbol{u} - \boldsymbol{c}_2^\top \boldsymbol{v})] + \mathrm{E}[R_2(\boldsymbol{u}, \boldsymbol{v})\boldsymbol{c}_2^\top \boldsymbol{K}_{(a)}\boldsymbol{u}(\boldsymbol{d}^\top \boldsymbol{u} - \boldsymbol{c}_2^\top \boldsymbol{v})] \\ =&\{2\mathrm{tr}\,(\boldsymbol{C}_c\boldsymbol{\Sigma}\boldsymbol{C}_d\boldsymbol{\Sigma}) + \kappa(\boldsymbol{C}_c, \boldsymbol{C}_d)\}\boldsymbol{c}_2^\top \boldsymbol{K}_{(a)}\boldsymbol{\Sigma}\boldsymbol{H}^\top(\boldsymbol{c}_1 - \boldsymbol{X}^\top \boldsymbol{K}^\top \boldsymbol{c}_2) + o(N). \end{aligned} \tag{3.18}$$

Since $(\boldsymbol{A}^{-1}\boldsymbol{E}_b\boldsymbol{A}^{-1})_{cd} = O(N^{-1})$, we can see that $I_{3ab} = o(N^{-1})$. From (B4), we have $I_{4a} = o(N^{-1})$. Thus, from (3.16), (3.17), and (3.18), we obtain the expression of $g_4(\boldsymbol{\psi})$ in (3.13). □

3.3 Evaluation of the MSE Under Normality

When the normal distributions are assumed for $\boldsymbol{v}$ and $\boldsymbol{\epsilon}$, we can decompose the MSE more directly. In this section, we assume that $\boldsymbol{v} \sim N_m(\boldsymbol{0}, \boldsymbol{R}_v(\boldsymbol{\psi}))$ and $\boldsymbol{\epsilon} \sim \mathrm{N}_N(\boldsymbol{0}, \boldsymbol{R}_e(\boldsymbol{\psi}))$. In the setup of normality, the conditional distribution of $\theta = \boldsymbol{c}_1^\top \boldsymbol{\beta} + \boldsymbol{c}_2^\top \boldsymbol{v}$ given $\boldsymbol{y}$ is

$$\theta \mid \boldsymbol{y} \sim \mathrm{N}\big(\widetilde{\theta}^{\mathrm{B}}(\boldsymbol{\psi}, \boldsymbol{\beta}), g_1(\boldsymbol{\psi})\big),$$

where $g_1(\boldsymbol{\psi})$ is given in (3.6) and $\widetilde{\theta}^{\mathrm{B}}(\boldsymbol{\psi}, \boldsymbol{\beta})$ is the Bayes estimator of θ given by

$$\widetilde{\theta}^{\mathrm{B}}(\boldsymbol{\psi}, \boldsymbol{\beta}) = \mathrm{E}[\theta \mid \boldsymbol{y}] = \boldsymbol{c}_1^\top \boldsymbol{\beta} + \boldsymbol{c}_2^\top \boldsymbol{R}_v \boldsymbol{Z}^\top \boldsymbol{\Sigma}^{-1}(\boldsymbol{y} - \boldsymbol{X}\boldsymbol{\beta}). \tag{3.19}$$

Substituting the GLS $\widetilde{\boldsymbol{\beta}}(\boldsymbol{\psi})$ into $\widetilde{\theta}^{\mathrm{B}}(\boldsymbol{\psi}, \boldsymbol{\beta})$ yields the BLUP is $\widetilde{\theta}^{\mathrm{BLUP}}(\boldsymbol{\psi})$, and the EBLUP $\widehat{\theta}^{\mathrm{EBLUP}}$ is given by substituting $\widehat{\boldsymbol{\psi}}$ in the BLUP. In this sense, the EBLUP is interpreted as the empirical Bayes (EB) estimator of θ. From the conditional expectation of θ given $\boldsymbol{y}$, the MSE of the EBLUP is decomposed as

$$\begin{aligned}\mathrm{MSE}(\boldsymbol{\psi}, \widehat{\theta}^{\mathrm{EBLUP}}) =& \mathrm{E}[\{\theta - \widetilde{\theta}^{\mathrm{B}}(\boldsymbol{\psi}, \boldsymbol{\beta})\}^2] + \mathrm{E}[\{\widehat{\theta}^{\mathrm{EBLUP}} - \widetilde{\theta}^{\mathrm{B}}(\boldsymbol{\psi}, \boldsymbol{\beta})\}^2] \qquad (3.20)\\ =& g_1(\boldsymbol{\psi}) + G_{\mathrm{B}}(\boldsymbol{\psi}), \qquad (3.21)\end{aligned}$$

where $G_{\mathrm{B}}(\boldsymbol{\psi}) = \mathrm{E}[\{\widehat{\theta}^{\mathrm{EBLUP}} - \widetilde{\theta}^{\mathrm{B}}(\boldsymbol{\psi}, \boldsymbol{\beta})\}^2]$.

For evaluating the term $G_{\mathrm{B}}(\boldsymbol{\psi})$, we assume the following conditions on $\widehat{\boldsymbol{\psi}}$.

(H1) $\widehat{\boldsymbol{\psi}} = \widehat{\boldsymbol{\psi}}(\boldsymbol{y})$ is an even function, namely, $\widehat{\boldsymbol{\psi}}(\boldsymbol{y}) = \widehat{\boldsymbol{\psi}}(-\boldsymbol{y})$.

(H2) $\widehat{\boldsymbol{\psi}}$ is a translation invariant function, namely, $\widehat{\boldsymbol{\psi}}(\boldsymbol{y} + \boldsymbol{X\alpha}) = \widehat{\boldsymbol{\psi}}(\boldsymbol{y})$ for any $\boldsymbol{\alpha} \in \mathbb{R}^p$.

When $\widehat{\boldsymbol{\psi}}$ satisfies (H1) and (H2), it can be shown that the term $G_{\mathrm{B}}(\boldsymbol{\psi})$ is decomposed as

$$\begin{aligned}G_{\mathrm{B}}(\boldsymbol{\psi}) =& \mathrm{E}[\{\widehat{\theta}^{\mathrm{BLUP}}(\boldsymbol{\psi}) - \widetilde{\theta}^{\mathrm{B}}(\boldsymbol{\psi}, \boldsymbol{\beta})\}^2] + \mathrm{E}[\{\widehat{\theta}^{\mathrm{EBLUP}} - \widetilde{\theta}^{\mathrm{BLUP}}(\boldsymbol{\psi})\}^2]\\ =& g_2(\boldsymbol{\psi}) + G_{\mathrm{E}}(\boldsymbol{\psi}),\end{aligned}$$

for $g_2(\boldsymbol{\psi})$ in (3.5). This means that the MSE can be decomposed as

$$\mathrm{MSE}(\boldsymbol{\psi}, \widehat{\theta}^{\mathrm{EBLUP}}) = g_1(\boldsymbol{\psi}) + g_2(\boldsymbol{\psi}) + G_{\mathrm{E}}(\boldsymbol{\psi}). \qquad (3.22)$$

Another approach to the decomposition under the normality is based on the following lemma.

Lemma 3.1 *Assume that conditions* (H1) *and* (H2) *hold. Let* $\boldsymbol{P} = \boldsymbol{I}_N - \boldsymbol{X}(\boldsymbol{X}^\top \boldsymbol{X})^{-1}\boldsymbol{X}^\top$. *Under the normality, the conditional distribution of* $\widehat{\theta}^{\mathrm{EBUP}} - \theta$ *given* $\boldsymbol{Py}$ *is*

$$\widehat{\theta}^{\mathrm{EBUP}} - \theta \mid \boldsymbol{Py} \sim \mathrm{N}\big(\widehat{\theta}^{\mathrm{EBUP}} - \widetilde{\theta}^{\mathrm{BLUP}}(\boldsymbol{\psi}), G(\boldsymbol{\psi})\big), \qquad (3.23)$$

for $G(\boldsymbol{\psi}) = g_1(\boldsymbol{\psi}) + g_2(\boldsymbol{\psi})$ *given in (3.4).*

This lemma was given in Diao et al. (2014) and used by Ito and Kubokawa (2021). Since $\widehat{\theta}^{\mathrm{EBUP}} - \widetilde{\theta}^{\mathrm{BLUP}}(\boldsymbol{\psi})$ is a function of $\boldsymbol{Py}$ under (H1) and (H2), from Lemma 3.1, the MSE of the EBLUP is decomposed as

$$\begin{aligned}\mathrm{MSE}(\boldsymbol{\psi}, \widehat{\theta}^{\mathrm{EBLUP}}) =& \mathrm{E}[\{\theta - \widetilde{\theta}^{\mathrm{BLUP}}(\boldsymbol{\psi})\}^2] + \mathrm{E}[\{\widehat{\theta}^{\mathrm{EBLUP}} - \widetilde{\theta}^{\mathrm{BLUP}}(\boldsymbol{\psi})\}^2]\\ =& G(\boldsymbol{\psi}) + G_{\mathrm{E}}(\boldsymbol{\psi}). \qquad (3.24)\end{aligned}$$

Since $G(\boldsymbol{\psi}) = g_1(\boldsymbol{\psi}) + g_2(\boldsymbol{\psi})$, one gets the same decomposition as in (3.22).

For evaluating $G_{\mathrm{E}}(\boldsymbol{\psi})$ asymptotically, we assume the following conditions.

(H3) $\widehat{\boldsymbol{\psi}}$ is $\sqrt{N}$-consistent, namely, $\widehat{\boldsymbol{\psi}} - \boldsymbol{\psi} = O_p(N^{-1/2})$.

(H4) $\boldsymbol{X}^\top \boldsymbol{X}$ is nonsingular and $\boldsymbol{X}^\top \boldsymbol{X}/N$ converges to a positive definite matrix.

Under these conditions, $G_{\mathrm{E}}(\boldsymbol{\psi})$ can be approximated as $G_{\mathrm{E}}(\boldsymbol{\psi}) = g_3(\boldsymbol{\psi}) + o(N^{-1})$, which leads to the second-order approximation of the MSE.

Theorem 3.3 *Assume that conditions* (H1)–(H4) *hold. Under the normality, the MSE is approximated as*

$$\mathrm{MSE}(\boldsymbol{\psi}, \widehat{\theta}^{\mathrm{EBLUP}}) = g_1(\boldsymbol{\psi}) + g_2(\boldsymbol{\psi}) + g_3(\boldsymbol{\psi}) + o(N^{-1}). \tag{3.25}$$

Finally, we note that such a decomposition as in (3.21) holds in the general parametric situation. Let $\widetilde{\theta}^{\mathrm{B}}(\boldsymbol{\psi})$ be the Bayes estimator of θ, namely, $\widetilde{\theta}^{\mathrm{B}}(\boldsymbol{\psi}) = \mathrm{E}[\theta \mid \boldsymbol{y}]$, where $\boldsymbol{\psi}$ is a hyperparameter. Let $\widehat{\boldsymbol{\psi}}$ be an estimator of $\boldsymbol{\psi}$. Then, the MSE of the empirical Bayes estimator $\widehat{\theta}^{\mathrm{EB}} = \widetilde{\theta}^{\mathrm{B}}(\widehat{\boldsymbol{\psi}})$ is decomposed as

$$\begin{aligned}\mathrm{MSE}(\boldsymbol{\psi}, \widehat{\theta}^{\mathrm{EB}}) =& \mathrm{E}[\{\theta - \widetilde{\theta}^{\mathrm{B}}(\boldsymbol{\psi})\}^2] + \mathrm{E}[\{\widehat{\theta}^{\mathrm{EB}} - \widetilde{\theta}^{\mathrm{B}}(\boldsymbol{\psi})\}^2] \\ =& \mathrm{E}[\mathrm{Var}(\theta \mid \boldsymbol{y})] + \mathrm{E}[\{\widehat{\theta}^{\mathrm{EB}} - \widetilde{\theta}^{\mathrm{B}}(\boldsymbol{\psi})\}^2],\end{aligned} \tag{3.26}$$

where $\mathrm{Var}(\theta \mid \boldsymbol{y})$ is the posterior (or conditional) variance of θ. In the case of normality, $\widehat{\theta}^{\mathrm{EB}}$ and $\mathrm{E}[\mathrm{Var}(\theta \mid \boldsymbol{y})]$ correspond to $\widehat{\theta}^{\mathrm{EBLUP}}$ and $g_1(\boldsymbol{\psi})$, respectively. It is noted that $\mathrm{Var}(\theta \mid \boldsymbol{y}) = O_p(1)$ and $\mathrm{E}[\{\widehat{\theta}^{\mathrm{EB}} - \widetilde{\theta}^{\mathrm{B}}(\boldsymbol{\psi})\}^2] = O(N^{-1})$ in many cases. The decomposition in (3.26) is used to obtain a second-order unbiased estimator of the MSE based on the bootstrap and jackknife methods.

3.4 Estimation of the MSE

For measuring the uncertainty of EBLUP, second-order unbiased estimators of MSE with bias of order $o(m^{-1})$ are widely adopted. Such asymptotically unbiased estimators can be derived by analytical methods, bootstrap methods, and jackknife methods. We here explain these three methods.

The analytical methods can be derived based on the second-order approximation of the MSE given in the previous section. This approach has been studied by Kackar and Harville (1984), Prasad and Rao (1990), Harville and Jeske (1992), Datta and Lahiri (2000), Datta et al. (2005), and Das et al. (2004). For some recent results including jackknife and bootstrap methods, see Lahiri and Rao (1995), Hall and Maiti (2006a), and Chen and Lahiri (2008).

The second-order approximation of the MSE is given in Theorem 3.1. It is noted that $g_1(\boldsymbol{\psi})$ is of order $O(1)$, while $g_2(\boldsymbol{\psi})$, $g_3(\boldsymbol{\psi})$, and $g_4(\boldsymbol{\psi})$ are of order $O(N^{-1})$. Since $g_1(\widehat{\boldsymbol{\psi}})$ has a second-order bias, the Taylor series expansion gives

$$\begin{aligned}g_1(\widehat{\boldsymbol{\psi}}) =& g_1(\boldsymbol{\psi}) + \sum_{a=1}^{q} g_{1(a)}(\boldsymbol{\psi})(\widehat{\psi}_a - \psi_a) \\ &+ \frac{1}{2}\sum_{a,b}^{q} g_{1(ab)}(\boldsymbol{\psi})(\widehat{\psi}_a - \psi_a)(\widehat{\psi}_b - \psi_b) + R_3^*(\boldsymbol{y}),\end{aligned} \tag{3.27}$$

where $R_3^*(\boldsymbol{y})$ is a remainder term. This leads to $\mathrm{E}[g_1(\widehat{\boldsymbol{\psi}})] = g_1(\boldsymbol{\psi}) + g_{12}(\boldsymbol{\psi}) + 2^{-1}g_{13}(\boldsymbol{\psi}) + \mathrm{E}[R_3^*(\boldsymbol{y})] + o(N^{-1})$, where

$$g_{12}(\boldsymbol{\psi}) = \sum_{a=1}^{q} g_{1(a)}(\boldsymbol{\psi})\mathrm{Bias}(\widehat{\psi}_a),$$

$$g_{13}(\boldsymbol{\psi}) = \sum_{a,b}^{q} g_{1(ab)}(\boldsymbol{\psi})\mathrm{Cov}(\widehat{\psi}_a, \widehat{\psi}_b).$$

Let

$$\begin{aligned} R_3(\boldsymbol{y}) =& \{g_2(\widehat{\boldsymbol{\psi}}) + g_3(\widehat{\boldsymbol{\psi}}) + 2g_4(\widehat{\boldsymbol{\psi}})\} - \{g_2(\boldsymbol{\psi}) + g_3(\boldsymbol{\psi}) + 2g_4(\boldsymbol{\psi})\} \\ &+ \{g_{12}(\widehat{\boldsymbol{\psi}}) + 2^{-1}g_{13}(\widehat{\boldsymbol{\psi}})\} - \{g_{12}(\boldsymbol{\psi}) + 2^{-1}g_{13}(\boldsymbol{\psi})\} + R_3^*(\boldsymbol{y}). \end{aligned}$$

If $\mathrm{E}[R_3(\boldsymbol{y})] = o(N^{-1})$, then $\mathrm{E}[g_1(\widehat{\boldsymbol{\psi}}) - g_{12}(\widehat{\boldsymbol{\psi}}) - 2^{-1}g_{13}(\widehat{\boldsymbol{\psi}})] = g_1(\boldsymbol{\psi}) + o(N^{-1})$. Since the second-order approximation of the MSE of EBLUP is $\mathrm{MSE}(\boldsymbol{\psi}, \widehat{\theta}^{\mathrm{EBLUP}}) = g_1(\boldsymbol{\psi}) + g_2(\boldsymbol{\psi}) + g_3(\boldsymbol{\psi}) + 2g_4(\boldsymbol{\psi}) + o(N^{-1})$, one gets a second-order unbiased estimator of the MSE.

Theorem 3.4 *Assume that conditions* (B1)–(B3) *hold. If* $\mathrm{E}[R_3(\boldsymbol{y})] = o(N^{-1})$, *then a second-order unbiased estimator of the MSE of EBLUP is*

$$\mathrm{mse}(\widehat{\theta}^{\mathrm{EBLUP}}) = g_1(\widehat{\boldsymbol{\psi}}) + g_2(\widehat{\boldsymbol{\psi}}) + g_3(\widehat{\boldsymbol{\psi}}) + 2g_4(\widehat{\boldsymbol{\psi}}) - g_{12}(\widehat{\boldsymbol{\psi}}) - 2^{-1}g_{13}(\widehat{\boldsymbol{\psi}}), \tag{3.28}$$

namely, $\mathrm{E}[\mathrm{mse}(\widehat{\theta}^{\mathrm{EBLUP}})] = \mathrm{MSE}(\widehat{\theta}^{\mathrm{EBLUP}}) + o(N^{-1})$.

The second-order unbiased estimate of the MSE can be given numerically by the bootstrap method. There are a few types of bootstrap methods for MSE. The most typical approach would be a hybrid bootstrap method given by Butar and Lahiri (2003). When $\boldsymbol{v}$ and $\boldsymbol{\epsilon}$ have normal distributions, the MSE is decomposed as (3.21) and we estimate $g_1(\boldsymbol{\psi})$ and $G_{\mathrm{B}}(\boldsymbol{\psi})$ via the parametric bootstrap. Here we explicitly write $\widetilde{\theta}^{\mathrm{BLUP}}(\boldsymbol{y}, \boldsymbol{\psi})$ instead of the estimator $\widetilde{\theta}^{\mathrm{BLUP}}(\boldsymbol{\psi})$ in order to address the dependence of the best predictor $\widetilde{\theta}^{\mathrm{BLUP}}(\boldsymbol{\psi})$ on the observation $\boldsymbol{y}$. Given the estimate $\widehat{\boldsymbol{\psi}} = \widehat{\boldsymbol{\psi}}(\boldsymbol{y})$ and $\widehat{\boldsymbol{\beta}} = \widehat{\boldsymbol{\beta}}(\boldsymbol{y})$, the parametric bootstrap method first generates the bootstrap sample $\boldsymbol{y}_{(b)}^*$ from the assumed model with $\widehat{\boldsymbol{\psi}}$ and $\widehat{\boldsymbol{\beta}}$ as the bth bootstrap sample, and then computes the bootstrap estimator $\widehat{\boldsymbol{\psi}}_b^* = \widehat{\boldsymbol{\psi}}^*(\boldsymbol{y}_{(b)}^*)$ of $\boldsymbol{\psi}$. Then, the hybrid bootstrap estimator is given by

$$2g_1(\widehat{\boldsymbol{\psi}}) - \frac{1}{B}\sum_{b=1}^{B} g_1(\widehat{\boldsymbol{\psi}}_b^*) + \frac{1}{B}\sum_{b=1}^{B}\left\{\widetilde{\theta}^{\mathrm{BLUP}}(\boldsymbol{y}_{(b)}^*, \widehat{\boldsymbol{\psi}}_b^*) - \widetilde{\theta}^{\mathrm{B}}(\boldsymbol{y}_{(b)}^*, \widehat{\boldsymbol{\psi}}, \widehat{\boldsymbol{\beta}})\right\}^2,$$

where B is the number of bootstrap replications. Another bootstrap method is the double bootstrap (e.g., Hall and Maiti 2006b) in which the naive bootstrap estimator is defined as

$$M(\boldsymbol{y}^*) = \frac{1}{B}\sum_{b=1}^{B}\left\{\widetilde{\theta}^{\mathrm{BLUP}}(\boldsymbol{y}^*_{(b)}, \widehat{\boldsymbol{\psi}}^*_b) - \theta^*_{(b)}\right\}^2,$$

where $\theta^*_{(b)}$ is a generated value of θ from the estimated model and $\boldsymbol{y}^*$ is a collection of the bootstrap sample. Then, the double bootstrap estimator computes C bootstrap MSE estimators $\{M(\boldsymbol{y}^*_1), \ldots, M(\boldsymbol{y}^*_C)\}$ and carry out bias correction. Compared with the hybrid bootstrap estimator, the double bootstrap method requires additional bootstrap replications, which could be computationally intensive in practice.

The jackknife method is also used to estimate the second-order unbiased estimate of the MSE. Jiang et al. (2002) suggested the use of the jackknife method for estimating g_1 and g_2, separately. Let $\widehat{\boldsymbol{\psi}}_{-\ell}$ denote the estimator of $\boldsymbol{\psi}$ based on all the data except for the ℓ-th area. Then, the jackknife estimator of MSE is given by $\widehat{g}_1^J + \widehat{g}_2^J$, where

$$\widehat{g}_1^J = g_1(\widehat{\boldsymbol{\psi}}) - \frac{m-1}{m}\sum_{\ell=1}^{m}\left\{g_1(\widehat{\boldsymbol{\psi}}_{-\ell}) - g_1(\widehat{\boldsymbol{\psi}})\right\},$$

$$\widehat{G}_{\mathrm{B}}^J = \frac{m-1}{m}\sum_{\ell=1}^{m}\left\{\widetilde{\theta}^{\mathrm{BLUP}}(\widehat{\boldsymbol{\psi}}_{-\ell}) - \widetilde{\theta}^{\mathrm{BLUP}}(\widehat{\boldsymbol{\psi}})\right\}^2.$$

Under some regularity conditions, it holds that the estimator is second-order unbiased.

Finally, we note that there is another type of MSE, called conditional MSE, defined as $\mathrm{E}[(\widehat{\theta}_i^{\mathrm{EBLUP}} - \theta_i)^2 \mid \boldsymbol{y}_i]$, which measures the estimation variability under given the observed data $\boldsymbol{y}_i$ of the ith area, where $\widehat{\theta}_i^{\mathrm{EBLUP}}$ is the EBLUP of θ_i for the ith area. The detailed investigation and comparisons with the standard (unconditional) MSE have been done in the literature (e.g., Datta et al. 2011; Torabi and Rao 2013). As noted in Booth and Hobert (1998) and Sugasawa and Kubokawa (2016), the difference between the conditional and unconditional MSEs under models based on normal distributions can be negligible under large sample sizes (i.e., large number of areas), whereas the difference is significant under non-normal distributions. Also unified jackknife methods for the conditional MSE are developed in Lohr and Rao (2009).

3.5 Confidence Intervals

Another approach to measuring uncertainty of EBLUP is a confidence interval based on EBLUP, and the confidence intervals which satisfy the nominal confidence level with second-order accuracy are desirable. There are mainly two methods for constructing the confidence intervals: the analytical method based on a Taylor series expansion and a parametric bootstrap method.

We derive a confidence interval of $\theta = \boldsymbol{c}_1^\top\boldsymbol{\beta} + \boldsymbol{c}_2^\top\boldsymbol{v}$ in the model (2.1) under the normality of $\boldsymbol{v}$ and $\boldsymbol{\epsilon}$. When $\boldsymbol{\psi}$ is known, from Lemma 3.1, a confidence interval of

θ with $100(1-\alpha)\%$ confidence coefficient is $\widetilde{\theta}^{\rm BLUP}(\boldsymbol{\psi}) \pm \sqrt{G(\boldsymbol{\psi})}z_{\alpha/2}$, where $z_{\alpha/2}$ is the $100(\alpha/2)\%$ upper quantile of the standard normal distribution. When $\boldsymbol{\psi}$ is unknown, we replace $\boldsymbol{\psi}$ with estimator $\widehat{\boldsymbol{\psi}}$ to get the naive confidence interval

$$CI_0 \,:\, \widehat{\theta}^{\rm EBLUP} \pm \sqrt{G(\widehat{\boldsymbol{\psi}})}z_{\alpha/2}, \tag{3.29}$$

which can be shown that the coverage probability tends to the nominal confidence coefficient $1-\alpha$ under conditions (H1)–(H4). Since $\mathrm{P}(\theta \in CI_0) = 1-\alpha+O(N^{-1})$, we want to derive a corrected confidence interval CI such that $\mathrm{P}(\theta \in CI) = 1-\alpha+o(N^{-1})$.

Define $B_1(\boldsymbol{\psi})$ and $B_2(\boldsymbol{\psi})$ by

$$\begin{aligned} B_1(\boldsymbol{\psi}) &= -\,(1/4)\mathrm{E}[\{g_1(\widehat{\boldsymbol{\psi}}) - g_1(\boldsymbol{\psi})\}^2]/\{G(\boldsymbol{\psi})\}^2 - 2g_3(\boldsymbol{\psi})/G(\boldsymbol{\psi}),\\ B_2(\boldsymbol{\psi}) &= -\,(3/4)\mathrm{E}[\{g_1(\widehat{\boldsymbol{\psi}}) - g_1(\boldsymbol{\psi})\}^2]/\{G(\boldsymbol{\psi})\}^2, \end{aligned}$$

for $g_3(\boldsymbol{\psi})$ given in (3.8).

Theorem 3.5 *Assume that conditions* (H1)–(H4) *hold. Also assume that* $\widehat{\boldsymbol{\psi}}$ *is second-order unbiased, namely,* $\mathrm{E}[\widehat{\boldsymbol{\psi}}] = \boldsymbol{\psi} + o(N^{-1})$. *Then,*

$$\mathrm{P}\Big[\frac{(\widehat{\theta}^{\rm EBLUP} - \theta)^2}{G(\widehat{\boldsymbol{\psi}})} \le x\Big] = F_1(x) + B_1 f_3(x) + B_2 f_5(x) + o(N^{-1}), \tag{3.30}$$

where $F_k(x)$ and $f_k(x)$ are the cumulative distribution and probability density functions of the chi-squared distribution with degrees of freedom k, respectively.

We omit the proof. For the details, see Diao et al. (2014) and Ito and Kubokawa (2021). Using Theorem 3.5, we can derive the Bartlett-type correction. For a function $h = h(\boldsymbol{\psi})$ with order $O(N^{-1})$, it is observed that

$$\mathrm{P}\Big[\frac{(\widehat{\theta}^{\rm EBLUP} - \theta)^2}{G(\widehat{\boldsymbol{\psi}})} \le x(1+h)\Big] = F_1(x) + hxf_1(x) + B_1 f_3(x) + B_2 f_5(x) + o(N^{-1}).$$

Note that $hxf_1(x) + B_1 f_3(x) + B_2 f_5(x)$ is of order $O(N^{-1})$. Thus, the second-order term vanishes if $hxf_1(x) = -B_1 f_3(x) - B_2 f_5(x) = 0$. Since $\Gamma(x+1) = x\Gamma(x)$ for the gamma function $\Gamma(x)$, the solution of this equation on h is

$$h^*(\boldsymbol{\psi}, x) = -B_1 + B_2 x/6.$$

Then, it holds that for any $x > 0$,

$$\mathrm{P}\Big[\frac{(\widehat{\theta}^{\rm EBLUP} - \theta)^2}{G(\widehat{\boldsymbol{\psi}})} \le x\{1 + h^*(\widehat{\boldsymbol{\psi}}, x)\}\Big] = F_1(x) + o(N^{-1}).$$

Hence, the corrected confidence region is given by

$$CI : \widehat{\theta}^{\rm EBLUP} \pm \sqrt{G(\widehat{\boldsymbol{\psi}})\{1 + h^*(\widehat{\boldsymbol{\psi}}, z_{\alpha/2}^2)\}} z_{\alpha/2}. \tag{3.31}$$

Similar corrected intervals were studied in Datta et al. (2002), Basu et al. (2003), Kubokawa (2010, 2011), Yoshimori and Lahiri (2014), and others.

The bootstrap method is used for constructing a confidence interval with second-order accuracy. We first provide a method using a pivotal statistic given in Chatterjee et al. (2008). Let $U(\boldsymbol{\psi}, \boldsymbol{\beta}) = (\theta - \widetilde{\theta}^{\rm B}(\boldsymbol{\psi}, \boldsymbol{\beta}))/\sqrt{g_1(\boldsymbol{\psi})}$ for the Bayes estimator $\widetilde{\theta}^{\rm B}(\boldsymbol{\psi}, \boldsymbol{\beta})$. Since $\theta \mid \boldsymbol{y} \sim {\rm N}(\widetilde{\theta}^{\rm B}(\boldsymbol{\psi}, \boldsymbol{\beta}), g_1(\boldsymbol{\psi}))$, it follows that $U(\boldsymbol{\psi}, \boldsymbol{\beta}) \sim {\rm N}(0, 1)$ when $\boldsymbol{\psi}$ is the true parameter. We approximate the distribution of $U(\widehat{\boldsymbol{\psi}}, \widehat{\boldsymbol{\beta}})$ via the parametric bootstrap, that is, we generate the parametric bootstrap sample $\theta_{(b)}^*$ as well as $\boldsymbol{y}_{(b)}^*$ from the estimated model and compute the bootstrap estimator $\widehat{\boldsymbol{\psi}}_{(b)}^*$ for $b = 1, \ldots, B$. Then the distribution of $U(\widehat{\boldsymbol{\psi}}, \widehat{\boldsymbol{\beta}})$ can be approximated by B bootstrap realizations $\{U_{(b)}^*,\ b = 1, \ldots, B\}$, where $U_{(b)}^* = (\theta_{(b)}^* - \widetilde{\theta}_{(b)}^{B*})/\sqrt{g_1(\widehat{\boldsymbol{\psi}}_{(b)}^*)}$. Letting $z_u^*(\alpha)$ and $z_l^*(\alpha)$ be the empirical upper and lower $100\alpha\%$ quantiles of the empirical distribution of $\{U_{(b)}^*,\ b = 1, \ldots, B\}$, the calibrated confidence interval is given by

$$\left[\widehat{\theta}^{\rm EBLUP} + z_l^*(\alpha/2)\{g_1(\widehat{\boldsymbol{\psi}})\}^{1/2},\ \widehat{\theta}^{\rm EBLUP} + z_u^*(\alpha/2)\{g_1(\widehat{\boldsymbol{\psi}})\}^{1/2}\right].$$

We next describe a general parametric bootstrap approach given in Hall and Maiti (2006b). Define $I_\alpha(\boldsymbol{\psi}) = (F_{\alpha/2}(\boldsymbol{\psi}), F_{1-\alpha/2}(\boldsymbol{\psi}))$ where $F_\alpha(\boldsymbol{\psi})$ is the α-quantile of the posterior distribution of θ, such as ${\rm N}(\widetilde{\theta}^{\rm B}(\boldsymbol{\psi}, \boldsymbol{\beta}), g_1(\boldsymbol{\psi}))$. Since the naive interval $I_\alpha(\widehat{\boldsymbol{\psi}})$ does not satisfy ${\rm P}(\theta_i \in I_\alpha(\widehat{\boldsymbol{\psi}})) = 1 - \alpha$, we calibrate a suitable α via the parametric bootstrap. Denote by $\widehat{I}_{\alpha(b)}^* = I_\alpha(\widehat{\boldsymbol{\psi}}_{(b)}^*)$ the bootstrap interval based on the bth bootstrap sample, and let $\widehat{\alpha}$ be the solution of the equation $B^{-1}\sum_{b=1}^{B} I(\theta_{i(b)}^* \in \widehat{I}_{\widehat{\alpha}(b)}^*) = \alpha$. Then, $I_{\widehat{\alpha}}(\widehat{\boldsymbol{\psi}})$ is the bootstrap-calibrated interval which has a coverage probability with second-order accuracy.

References

Basu R, Ghosh JK, Mukerjee R (2003) Empirical Bayes prediction intervals in a normal regression model: higher order asymptotics. Statist. Prob. Lett. 63:197–203

Booth JS, Hobert P (1998) Standard errors of prediction in generalized linear mixed models. J. Am. Statist. Assoc. 93:262–271

Buttar FB, Lahiri P (2003) On measures of uncertainty of empirical Bayes small-area estimators. J. Statist. Plann. Inf. 112:63–76

Chatterjee S, Lahiri P, Li H (2008) Parametric bootstrap approximation to the distribution of EBLUP and related prediction intervals in linear mixed models. Ann. Statist. 36:1221–1245

Chen S, Lahiri P (2008) On mean squared prediction error estimation in small area estimation problems. Commun Statist-Theory Methods 37: 1792–1798

Das K, Jiang J, Rao JNK (2004) Mean squared error of empirical predictor. Ann. Statist. 32:818–840
Datta GS, Ghosh M, Smith DD, Lahiri P (2002) On an asymptotic theory of conditional and unconditional coverage probabilities of empirical Bayes confidence Intervals. Scandinavian J. Statist. 29:139–152
Datta GS, Hall P, Mandal A (2011) Model selection by testing for the presence of small-area effects, and application to area-level data. J. Am. Statist. Assoc. 106:362–374
Datta GS, Lahiri P (2000) A unified measure of uncertainty of estimated best linear unbiased predictors in small area estimation problems. Statist. Sinica 10:613–627
Datta GS, Rao JNK, Smith DD (2005) On measuring the variability of small area estimators under a basic area level model. Biometrika 92:183–196
Diao L, Simith DD, Datta GS, Maiti T, Opsomer JD (2014) Accurate confidence interval estimation of small area parameters under the Fay-Herriot model. Scand. J. Statist. 41:497–515
Ghosh M, Maiti T (2004) Small-area estimation based on natural exponential family quadratic variance function models and survey weights. Biometrika 91:95–112
Hall P, Maiti T (2006) Nonparametric estimation of mean-squared prediction error in nested-error regression models. Ann. Statist. 34:1733–1750
Hall P, Maiti T (2006) On parametric bootstrap methods for small area prediction. J. R. Statist. Soc. 68:221–238
Harville DA, Jeske DR (1992) Mean squared error of estimation or prediction under a general linear model. J. Am. Statist. Assoc. 87:724–731
Ito T, Kubokawa T (2021) Corrected empirical Bayes confidence region in a multivariate Fay-Herriot model. J. Statist. Plann. Inf. 211:12–32
Jiang J, Lahiri P, Wan SM (2002) A unified Jackknife theory for empirical best prediction with m-estimation. Ann. Statist. 30:1782–1810
Kackar RN, Harville DA (1984) Approximations for standard errors of estimators of fixed and random effects in mixed linear models. J. Am. Statist. Assoc. 79:853–862
Kubokawa T (2010) Corrected empirical Bayes confidence intervals in nested error regression models. J. Korean Statist. Soc. 39:221–236
Kubokawa T (2011) On measuring uncertainty of small area estimators with higher order accuracy. J. Japan Statist. Soc. 41:93–119
Lahiri P, Rao JNK (1995) Robust estimation of mean squared error of small area estimators. J. Am. Statist. Assoc. 90:758–766
Lohr SL, Rao JNK (2009) Jackknife estimation of mean squared error of small area predictors in nonlinear mixed model. Biometrika 96:457–468
Prasad NGN, Rao JNK (1990) The estimation of the mean squared error of small-area estimators. J. Am. Statist. Assoc. 85:163–171
Sugasawa S, Kubokawa T (2016) On conditional prediction errors in mixed models with application to small area estimation. J. Multivariate Anal. 148:18–33
Torabi M, Rao JNK (2013) Estimation of mean squared error of model-based estimators of small area means under a nested error linear regression model. J. Multivariate Anal. 117:76–87
Yoshimori M, Lahiri P (2014) A second-order efficient empirical Bayes confidence interval. Ann. Statist. 42:1233–1261

Chapter 4
Basic Mixed-Effects Models for Small Area Estimation

Statistical inference in the general linear mixed models is explained in the previous chapters. As basic models used in small area estimation, in this chapter, we treat two most standard models, known as the Fay–Herriot model and the nested error regression model, which have been extensively used for analyzing area-level and unit-level regional or spatial data, respectively. We here provide estimators of variance components, EBLUP of area means, and their asymptotic properties in these specific models.

4.1 Basic Area-Level Model

4.1.1 Fay–Herriot Model

Most public data are reported based on accumulated data like sample means from counties and cities. The Fay–Herriot (FH) model introduced by Fay and Herriot (1979) is the mixed model for estimating the true areal means $\theta_1, \ldots, \theta_m$ based on area-level summary statistics denoted by $y_1, \ldots, y_m$, where y_i is called a direct estimate of θ_i for $i = 1, \ldots, m$. Note that y_i is a crude estimator θ_i with large variance, because the sample size for calculating y_i could be small in practice. Let $\boldsymbol{x}_i$ be a vector of known area characteristics with an intercept term. The FH model is written as

$$y_i = \theta_i + \varepsilon_i, \quad \theta_i = \boldsymbol{x}_i^\top \boldsymbol{\beta} + v_i, \quad i = 1, \ldots, m, \tag{4.1}$$

where $\boldsymbol{\beta}$ is a p-variate vector of regression coefficients, ε_i and v_i are the sampling errors and the random effects, respectively, and are independently distributed as $\mathrm{E}[\varepsilon_i] = \mathrm{E}[v_i] = 0$, $\mathrm{Var}(\varepsilon_i) = D_i$, $\mathrm{Var}(v_i) = \psi$, $\mathrm{E}[\varepsilon_i^4] = (\kappa_e + 3)D_i^2$ and $\mathrm{E}[v_i^4] =$

S. Sugasawa and T. Kubokawa, *Mixed-Effects Models and Small Area Estimation*, JSS Research Series in Statistics, https://doi.org/10.1007/978-981-19-9486-9_4

$(\kappa_v + 3)\psi^2$. Here, D_i is a variance of y_i given θ_i, which is assumed to be known, and ψ is an unknown variance parameter. Also, κ_e is known, while κ_v is unknown. The assumption that D_i and κ_e are known seems restrictive, because they can be estimated from data a priori. This issue will be addressed in Chap. 8.

The Fay–Herriot model belongs to the general mixed-effects models treated in Chap. 2 with the correspondences $N = m$, $\boldsymbol{Z} = \boldsymbol{I}_m$, $\boldsymbol{R}_v = \psi \boldsymbol{I}_m$, $\boldsymbol{R}_e = \boldsymbol{D} = \mathrm{diag}\,(D_1, \ldots, D_m)$, $\boldsymbol{\Sigma} = \psi \boldsymbol{I}_m + \boldsymbol{D}$, $\boldsymbol{y} = (y_1, \ldots, y_m)^\top$, and $\boldsymbol{X} = (\boldsymbol{x}_1, \ldots, \boldsymbol{x}_m)^\top$. It is noted that $\theta_i = \boldsymbol{x}_i^\top \boldsymbol{\beta} + v_i$ corresponds to the case of $\boldsymbol{c}_1 = \boldsymbol{x}_i$ and $\boldsymbol{c}_2 = \boldsymbol{e}_i$ in (2.10), where the j-th element of $\boldsymbol{e}_i$ is one for $j = i$ and zero for $j \neq i$. From Theorem 2.1, the best linear unbiased predictor (BLUP) of θ_i is

$$\widetilde{\theta}_i^{\mathrm{BLUP}}(\psi) = \boldsymbol{x}_i^\top \widetilde{\boldsymbol{\beta}}(\psi) + \gamma_i (y_i - \boldsymbol{x}_i^\top \widetilde{\boldsymbol{\beta}}(\psi)) = y_i - (1 - \gamma_i)(y_i - \boldsymbol{x}_i^\top \widetilde{\boldsymbol{\beta}}(\psi)), \quad (4.2)$$

where $\gamma_i = \gamma_i(\psi) = \psi/(\psi + D_i)$ and $\widetilde{\boldsymbol{\beta}}(\psi)$ is the generalized least squares estimator (GLS) of $\boldsymbol{\beta}$ given by

$$\widetilde{\boldsymbol{\beta}}(\psi) = (\boldsymbol{X}^\top \boldsymbol{\Sigma}^{-1} \boldsymbol{X})^{-1} \boldsymbol{X}^\top \boldsymbol{\Sigma}^{-1} \boldsymbol{y} = \Big(\sum_{i=1}^m \gamma_i \boldsymbol{x}_i \boldsymbol{x}_i^\top \Big)^{-1} \Big(\sum_{i=1}^m \gamma_i \boldsymbol{x}_i y_i \Big).$$

In practice, the random effects variance ψ is unknown, and should be replaced in γ_i and $\widetilde{\boldsymbol{\beta}}(\psi)$ by sample estimate $\widehat{\psi}$, which yields the empirical BLUP (EBLUP)

$$\widehat{\theta}_i^{\mathrm{EBLUP}} = \boldsymbol{x}_i^\top \widehat{\boldsymbol{\beta}} + \widehat{\gamma}_i (y_i - \boldsymbol{x}_i^\top \widehat{\boldsymbol{\beta}}), \quad \widehat{\boldsymbol{\beta}} = \widetilde{\boldsymbol{\beta}}(\widehat{\psi}), \ \widehat{\gamma}_i = \widehat{\psi}/(\widehat{\psi} + D_i),$$

in the frequentist's framework, or the empirical Bayes estimator in the Bayesian framework. A standard way to estimate ψ is the maximum likelihood estimator based on the marginal distribution of y_is. Other options would be the restricted maximum likelihood estimator and moment-type estimators as considered in Fay–Herriot (1979) and Prasad–Rao (1990). We will revisit this issue in Sect. 4.1.2. Alternatively, we may employ the hierarchical Bayes (HB) approach by assigning prior distributions on unknown parameters $\boldsymbol{\beta}$ and ψ, and compute a posterior distribution of θ_i, which produces the point estimator as well as credible intervals. Due to the recent advancement of computational techniques, the HB approaches now became standard in this context (Rao-Molina, 2015).

Since $\widehat{\boldsymbol{\beta}}$ is constructed based on all the data, the regression estimator $\boldsymbol{x}_i^\top \widehat{\boldsymbol{\beta}}$ would be much more stable than the direct estimator y_i. Then, the EBLUP can be interpreted as a shrinkage estimator that shrinks the unstable direct estimator y_i toward the stable estimator $\boldsymbol{x}_i^\top \widehat{\boldsymbol{\beta}}$, depending on the shrinkage coefficient $\widehat{\gamma}_i$. Note that if D_i is large compared with $\widehat{\psi}$, which means that y_i has a large fluctuation, $\widehat{\gamma}_i$ is small, so that y_i is more shrunk toward $\boldsymbol{x}_i^\top \widehat{\boldsymbol{\beta}}$, and vise versa. Such desirable properties of EBLUP come from the structure of the linear mixed model described as (observation) = (common mean) + (random effect) + (error term).

[1] Shrinkage via random effects. In the case that v_i is a fixed parameter, the best estimator of θ_i is y_i. When v_i is a random effect, however, the covariance matrix of (y_i, v_i) is

$$\begin{pmatrix} \mathrm{Var}(y_i) & \mathrm{Cov}(y_i, v_i) \\ \mathrm{Cov}(y_i, v_i) & \mathrm{Var}(v_i) \end{pmatrix} = \begin{pmatrix} \psi + D_i & \psi \\ \psi & \psi \end{pmatrix},$$

namely, the correlation yields between y_i and v_i. From this correlation, it follows that the conditional expectation under normality is written as $\mathrm{E}[v_i|y_i] = \psi(\psi + D_i)^{-1}(y_i - \boldsymbol{x}_i^\top \boldsymbol{\beta})$, which means that the conditional expectation shrinks y_i towards $\boldsymbol{x}_i^\top \boldsymbol{\beta}$. Thus, the random effect v_i produces the function of shrinkage in EBLUP.

[2] Pooling data via common parameters. The regression coefficient $\boldsymbol{\beta}$ is embedded as a common parameter in all the small areas. This means that all the data are used for estimating the common parameter, which results in the pooling effect. Thus, the setup via the common parameters leads to the pooling effect, and one gets the stable estimator $\boldsymbol{x}_i^\top \widehat{\boldsymbol{\beta}}$.

As stated above, we can obtain stable estimates via pooling data through restricting parameters to some constraints like equality or inequality, and we can shrink y_i toward the stable estimates through incorporating random effects. This enables us to boost up the precision of the prediction. As seen from the fact that EBLUP is interpreted as the empirical Bayes estimator, this perspective was recognized by Efron and Morris (1975) in the context of the empirical Bayes method, and the usefulness of the Bayesian methods may be based on such perspective.

[3] Henderson's EBLUP and Stein's shrinkage. Consider the case of $D_1 = \cdots = D_m = \sigma_0^2$. When ψ is estimated by $\widehat{\psi}^{\mathrm{ST}} = \sum_{j=1}^m (y_j - \boldsymbol{x}_j^\top \widehat{\boldsymbol{\beta}})^2/(m - p - 2) - \sigma_0^2 = \|\boldsymbol{y} - \boldsymbol{X}\widehat{\boldsymbol{\beta}}\|^2/(m - p - 2) - \sigma_0^2$, from (4.2), the EBLUP is

$$\widehat{\theta}_i^{\mathrm{ST}} = y_i - \frac{(m - p - 2)\sigma_0^2}{\|\boldsymbol{y} - \boldsymbol{X}\widehat{\boldsymbol{\beta}}\|^2}(y_i - \boldsymbol{x}_i^\top \widehat{\boldsymbol{\beta}}).$$

This is the Stein estimator suggested by Stein (1956) and James–Stein (1961), who proved that $\widehat{\theta}_i^{\mathrm{ST}}$ has a uniformly smaller risk than y_i in the framework of simultaneous estimation of $(\theta_1, \ldots, \theta_m)$ under normality if $m - p \geq 3$. This fact implies that EBLUP has a larger precision than the crude estimate y_i. It is interesting to note that a similar concept came out at the same time by Henderson (1950) for practical use and Stein (1956) for theoretical interest.

4.1.2 Asymptotic Properties of EBLUP

We provide asymptotic properties of estimators of ψ and the EBLUP in the Fay–Herriot model. The estimators we treat for ψ are given in (2.20), namely $\widehat{\psi} = \max(\psi^*, 0)$ for the solution of the equation

$$\ell = \ell(\psi) = \boldsymbol{y}^\top (\boldsymbol{I} - \boldsymbol{X}\boldsymbol{L})^\top \boldsymbol{W}(\boldsymbol{I} - \boldsymbol{X}\boldsymbol{L})\boldsymbol{y} - \mathrm{tr}\,\{(\boldsymbol{I} - \boldsymbol{X}\boldsymbol{L})^\top \boldsymbol{W}(\boldsymbol{I} - \boldsymbol{X}\boldsymbol{L})\boldsymbol{\Sigma}\} = 0, \tag{4.3}$$

for $\boldsymbol{\Sigma} = \psi \boldsymbol{I}_m + \boldsymbol{D}$, where $\boldsymbol{W} = \boldsymbol{W}(\psi)$ is an $m \times m$ diagonal matrix of twice continuously differentiable functions of ψ, and $\boldsymbol{L} = \boldsymbol{L}(\psi)$ is a $p \times m$ matrix of twice continuously differentiable functions of ψ satisfying $\boldsymbol{L}\boldsymbol{X} = \boldsymbol{I}_m$. We assume the following conditions.

(FH1) D_is satisfy that $D_L < D_i < D_U$ for $0 < D_L \le D_U < \infty$.

(FH2) $\boldsymbol{X}$ is of full rank and $(\boldsymbol{X}\boldsymbol{L})_{ij} = O(m^{-1})$.

(FH3) $\mathrm{E}[v_i^{10}] < \infty$ and $\mathrm{E}[\varepsilon_i^{10}] < \infty$ for some $\delta > 0$.

Then, it can be verified that (C1), (C2), and (C5) are satisfied. Conditions (C3) and (C4) are replaced with the following.

(C3†) $\mathrm{E}[\{\ell_{(1)}^\dagger + \mathrm{tr}\,(\boldsymbol{W})\}^2(\widehat{\psi} - \psi)^2] = o(m)$ for $\ell_{(1)}^\dagger = \ell_{(1)}(\psi^\dagger)$, where $\psi^\dagger$ is on the line segment between ψ and $\widehat{\psi}$.

(C4†) The expectation of the remainder term $\mathrm{E}[r(\boldsymbol{y})]$ is of order $o(m^{-1})$, where

$$\begin{aligned} r(\boldsymbol{y}) = & \frac{1}{\mathrm{tr}\,(\boldsymbol{W})}\{\ell_{(1)} + \mathrm{tr}\,(\boldsymbol{W})\}\Big\{\widehat{\psi} - \psi - \frac{\ell}{\mathrm{tr}\,(\boldsymbol{W})}\Big\} - \frac{2\mathrm{tr}\,(\boldsymbol{W}_{(1)})}{\mathrm{tr}\,(\boldsymbol{W})}\Big\{\widehat{\psi} - \psi - \frac{\ell}{\mathrm{tr}\,(\boldsymbol{W})}\Big\}\frac{\ell}{\mathrm{tr}\,(\boldsymbol{W})} \\ & - \frac{2\mathrm{tr}\,(\boldsymbol{W}_{(1)})}{\mathrm{tr}\,(\boldsymbol{W})}\Big\{\widehat{\psi} - \psi - \frac{\ell}{\mathrm{tr}\,(\boldsymbol{W})}\Big\}^2 + \frac{(\widehat{\psi} - \psi)^2}{2\mathrm{tr}\,(\boldsymbol{W})}\{\ell_{(1,1)}^* + 2\mathrm{tr}\,(\boldsymbol{W}_{(1)})\}, \end{aligned} \tag{4.4}$$

for $\ell_{(1,1)}^* = (\partial^2 \ell(\psi)/\partial\psi^2)|_{\psi=\psi^*}$, where ψ^* is on the line segment between ψ and $\widehat{\psi}$.

These conditions involve $\ell(\psi)$ and the derivatives. In the case of $\boldsymbol{L} = (\boldsymbol{X}^\top\boldsymbol{X})^{-1}\boldsymbol{X}^\top$, we have $\ell = \boldsymbol{u}^\top \boldsymbol{P}\boldsymbol{W}\boldsymbol{P}\boldsymbol{u} - \mathrm{tr}\,(\boldsymbol{W}\boldsymbol{P}\boldsymbol{\Sigma}\boldsymbol{P})$, $\ell_{(1)} = \boldsymbol{u}^\top \boldsymbol{P}\boldsymbol{W}_{(1)}\boldsymbol{P}\boldsymbol{u} - \mathrm{tr}\,(\boldsymbol{W}_{(1)}\boldsymbol{P}\boldsymbol{\Sigma}\boldsymbol{P}) - \mathrm{tr}\,(\boldsymbol{W}\boldsymbol{P})$ and $\ell_{(1,1)} = \boldsymbol{u}^\top \boldsymbol{P}\boldsymbol{W}_{(1,1)}\boldsymbol{P}\boldsymbol{u} - \mathrm{tr}\,(\boldsymbol{W}_{(1,1)}\boldsymbol{P}\boldsymbol{\Sigma}\boldsymbol{P}) - 2\mathrm{tr}\,(\boldsymbol{W}_{(1)}\boldsymbol{P})$ for $\boldsymbol{u} = \boldsymbol{y} - \boldsymbol{X}\boldsymbol{\beta}$ and $\boldsymbol{P} = \boldsymbol{I}_m - \boldsymbol{X}(\boldsymbol{X}^\top\boldsymbol{X})^{-1}\boldsymbol{X}^\top$.

From Theorem 2.5 and Proposition 2.1, we have the following theorem.

Theorem 4.1 *Let $\boldsymbol{W} = \boldsymbol{W}(\psi)$ be an $m \times m$ diagonal matrix of twice continuously differentiable functions of ψ. Let $\boldsymbol{L} = \boldsymbol{L}(\psi)$ be a $p \times m$ matrix of twice continuously differentiable functions of ψ satisfying $\boldsymbol{L}\boldsymbol{X} = \boldsymbol{I}_m$. Assume that conditions* (FH1)-(FH3), (C3†), *and* (C4†) *hold. Then, the variance and the bias of $\widehat{\psi}$ is*

$$\mathrm{Var}(\widehat{\psi}) = \frac{1}{\{\mathrm{tr}\,(\boldsymbol{W})\}^2}\{2\mathrm{tr}\,(\boldsymbol{W}\boldsymbol{\Sigma}\boldsymbol{W}\boldsymbol{\Sigma}) + \kappa(\boldsymbol{W}, \boldsymbol{W})\} + o(m^{-1}), \tag{4.5}$$

$$\begin{aligned} \mathrm{Bias}(\widehat{\psi}) = & \frac{1}{\{\mathrm{tr}\,(\boldsymbol{W})\}^2}\Big\{2\mathrm{tr}\,(\boldsymbol{W}_{(1)}\boldsymbol{\Sigma}\boldsymbol{W}\boldsymbol{\Sigma}) + \kappa(\boldsymbol{W}_{(1)}, \boldsymbol{W}) \\ & - \frac{\mathrm{tr}\,(\boldsymbol{W}_{(1)})}{\mathrm{tr}\,(\boldsymbol{W})}\{2\mathrm{tr}\,(\boldsymbol{W}\boldsymbol{\Sigma}\boldsymbol{W}\boldsymbol{\Sigma}) + \kappa(\boldsymbol{W}, \boldsymbol{W})\}\Big\} + o(m^{-1}), \end{aligned} \tag{4.6}$$

where for diagonal matrices $\boldsymbol{A}_0$ and $\boldsymbol{B}_0$,

$$\kappa(\boldsymbol{A}_0, \boldsymbol{B}_0) = \kappa_e \mathrm{tr}\,(\boldsymbol{D}^2\boldsymbol{A}_0\boldsymbol{B}_0) + \psi^2\kappa_v \mathrm{tr}\,(\boldsymbol{A}_0\boldsymbol{B}_0).$$

Concerning the MSE of EBLUP $\widehat{\theta}_i^{\rm EBLUP}$ of $\theta_i = \boldsymbol{x}_i^\top \boldsymbol{\beta} + v_i$, the second-order approximation is given from Theorem 3.1. Note that the nonations given in (3.2) correspond to $\boldsymbol{K} = \psi \boldsymbol{\Sigma}^{-1}$, $\boldsymbol{H} = (\boldsymbol{X}^\top \boldsymbol{\Sigma}^{-1} \boldsymbol{X})^{-1} \boldsymbol{X}^\top \boldsymbol{\Sigma}^{-1}$, $\boldsymbol{d}^\top \boldsymbol{u} = \gamma_i u_i + (1-\gamma_i)\boldsymbol{x}_i^\top \boldsymbol{H}\boldsymbol{u}$ and $\boldsymbol{d}_{(1)}^\top \boldsymbol{u} = D_i^{-1}(1-\gamma_i)^2 u_i + \boldsymbol{m}_1^\top \boldsymbol{u}$ for $\boldsymbol{m}_1^\top = (1-\gamma_i)\boldsymbol{x}_i^\top \boldsymbol{H}_{(a)} - D_i^{-1}(1-\gamma_i)^2 \boldsymbol{x}_i^\top \boldsymbol{H}$. Thus from (3.5),

$$g_1(\psi) = D_i \gamma_i, \quad g_2(\psi) = (1-\gamma_i)^2 \boldsymbol{x}_i^\top (\boldsymbol{X}^\top \boldsymbol{\Sigma}^{-1} \boldsymbol{X})^{-1} \boldsymbol{x}_i. \tag{4.7}$$

Conditions (B1)–(B3) can be guaranteed under (FH1)–(FH3) and the following condition.

(FH4) $\mathrm{E}[(\widehat{\psi}-\psi)^2] = O(m^{-1})$, $\mathrm{E}[(\widehat{\psi}-\psi)^4 u_i v_i] = o(m^{-1})$, $\mathrm{E}[(\widehat{\psi}-\psi)^4 u_i u_j] = o(m^{-1})$, $\mathrm{E}[u_i u_j R_0(u_i, u_j)] = o(m^{-1})$ and $\mathrm{E}[(u_i^2 + u_i v_i) R_0(v_i, \varepsilon_i)] = o(m^{-1})$ for $R_0(u_i, u_j) = \mathrm{E}[(\widehat{\psi}-\psi)^2 \mid u_i, u_j] - \mathrm{E}[(\widehat{\psi}-\psi)^2]$ for $i, j \in \{1, \ldots, m\}$. For $i \in \{1, \ldots, m\}$, $\mathrm{E}[(\widehat{\psi}-\psi)(u_i + v_i)\sum_{j=1}^m c_j u_j = o(m^{-1})$ for constants c_j satisfying $c_j = O(m^{-1})$.

Theorem 4.2 *Under conditions* (FH1)–(FH4)*, the MSE of* $\widehat{\theta}^{\rm EBLUP}$ *is approximated as*

$$\mathrm{MSE}(\psi, \widehat{\theta}^{\rm EBLUP}) = g_1(\psi) + g_2(\psi) + g_3(\psi) + 2g_4(\psi) + o(m^{-1}), \tag{4.8}$$

where $g_1(\psi)$ *and* $g_2(\psi)$ *are given in (4.7), and* $g_3(\psi)$ *and* $g_4(\psi)$ *are*

$$g_3(\psi) = D_i^{-1}(1-\gamma_i)^3 \mathrm{Var}(\widehat{\psi}), \tag{4.9}$$

$$g_4(\psi) = D_i^{-1}(1-\gamma_i)^2 \mathrm{E}[\widehat{\psi} u_i(\gamma_i u_i - v_i)]. \tag{4.10}$$

Under the normality of $\boldsymbol{v}$ *and* $\boldsymbol{\epsilon}$, *it holds that* $g_4(\psi) = 0$.

The proof of Theorem 4.2 is given in the end of this subsection.

The analytical method for a second-order unbiased estimator of the MSE of EBLUP can be derived by Theorem 3.4. Since $g_1(\widehat{\psi})$ has a second-order bias, the Taylor series expansion in (3.27) is written as

$$g_1(\widehat{\psi}) = g_1(\psi) + (1-\gamma_i)^2(\widehat{\psi}-\psi) - \frac{(1-\gamma_i)^3}{D_i}(\widehat{\psi}-\psi)^2 + \frac{(1-\gamma_i^*)^4}{D_i^2}(\widehat{\psi}-\psi)^3,$$

which leads to

$$\begin{aligned} \mathrm{E}[g_1(\widehat{\psi})] = {} & g_1(\psi) + (1-\gamma_i)^2 \mathrm{Bias}(\widehat{\psi}) - (1-\gamma_i)^3 D_i^{-1}\{\mathrm{Var}(\widehat{\psi}) + (\mathrm{Bias}(\widehat{\psi}))^2\} \\ & + \mathrm{E}[(1-\gamma_i^*)^4 D_i^{-2}(\widehat{\psi}-\psi)^3], \end{aligned}$$

where $\gamma_i^* = \gamma_i(\psi^*)$ for a point $\widehat{\psi}^*$ on the line segment between ψ and $\widehat{\psi}$. Note that $(1-\gamma_i^*)^4 \le 1$. Let

$$
\begin{aligned}
R_3^{\rm FH}(\boldsymbol{y}) = & \{(1-\widehat{\gamma}_i)^2\widehat{\rm Bias}(\psi) - (1-\gamma_i)^2{\rm Bias}(\widehat{\psi})\} + (\widehat{\psi}-\psi)^3 \\
& + \{g_2(\widehat{\psi}) - g_2(\psi)\} + \{g_3(\widehat{\psi}) - g_3(\psi)\} + \{g_4(\widehat{\psi}) - g_4(\psi)\},
\end{aligned}
$$

where $\widehat{\gamma}_i = \gamma_i(\widehat{\psi})$, and $\widehat{\rm Bias}(\psi)$ and $\widehat{\rm Var}(\psi)$ are the plug-in estimators of ${\rm Bias}(\widehat{\psi})$ and ${\rm Var}(\widehat{\psi})$.

Theorem 4.3 *Assume that conditions* (FH1)–(FH4) *hold. If* ${\rm E}[R_3^{\rm FH}(\boldsymbol{y})] = o(m^{-1})$, *then a second-order unbiased estimator of the MSE of EBLUP is*

$$
{\rm mse}(\widehat{\theta}^{\rm EBLUP}) = g_1(\widehat{\psi}) + g_2(\widehat{\psi}) + 2g_3(\widehat{\psi}) + 2g_4(\widehat{\psi}) - (1-\widehat{\gamma}_i)^2\widehat{\rm Bias}(\psi), \quad (4.11)
$$

namely, ${\rm E}[{\rm mse}(\widehat{\theta}^{\rm EBLUP})] = {\rm MSE}(\widehat{\theta}^{\rm EBLUP}) + o(m^{-1})$.

For the estimator $\widehat{\psi}$ given in (4.3), we can derive more specific formulae when adding the following condition.

(FH5) ${\rm E}[r(\boldsymbol{y})u_i\{\gamma_i u_i - v_i + (1-\gamma_i)\boldsymbol{x}_i^\top \boldsymbol{H}\boldsymbol{u}\}] = o(m^{-1})$ for $r(\boldsymbol{y})$ given in (4.4). ${\rm E}[R_1^{\rm FH}(u_i, v_i)u_i\{\gamma_i u_i - v_i + (1-\gamma_i)\boldsymbol{x}_i^\top \boldsymbol{H}\boldsymbol{u}\}] = o(m)$ and ${\rm E}[R_2^{\rm FH}(u_i, v_i)u_i\{\gamma_i u_i - v_i + (1-\gamma_i)\boldsymbol{x}_i^\top \boldsymbol{H}\boldsymbol{u}\}] = o(m)$, where

$$
\begin{aligned}
R_1^{\rm FH}(u_i, v_i) &= {\rm E}[\{\ell_{(1)} + {\rm tr}\,(\boldsymbol{W})\}\ell \mid u_i, v_i] - {\rm E}[\ell_{(1)}\ell], \\
R_2^{\rm FH}(u_i, v_i) &= {\rm E}[\ell^2 \mid u_i, v_i] - {\rm E}[\ell^2].
\end{aligned}
$$

Theorem 4.4 *Assume that conditions* (FH1)-(FH5), (C3†), *and* (C4†) *hold. For the estimator $\widehat{\psi}$ given in (4.3), the functions $g_3(\psi)$ and $g_4(\psi)$ are written as*

$$
g_3(\psi) = \frac{D_i^{-1}(1-\gamma_i)^3}{\{{\rm tr}\,(\boldsymbol{W})\}^2}\{2{\rm tr}\,(\boldsymbol{W\Sigma W\Sigma}) + \kappa(\boldsymbol{W}, \boldsymbol{W})\} + o(m^{-1}), \quad (4.12)
$$

$$
g_4(\psi) = \frac{(\boldsymbol{W})_{ii}D_i\gamma_i(1-\gamma_i)}{{\rm tr}\,(\boldsymbol{W})}\left\{\kappa_e(1-\gamma_i) - \kappa_v\gamma_i\right\}. \quad (4.13)
$$

We treat the three specific estimators of ψ, namely, the Prasad–Rao estimator, the Fay–Herriot estimator and the REML estimator which correspond to $\boldsymbol{W}^{\rm PR} = \boldsymbol{I}_m$, $\boldsymbol{W}^{\rm RE} = \boldsymbol{\Sigma}^{-2}$ and $\boldsymbol{W}^{\rm FH} = \boldsymbol{\Sigma}^{-1}$, respectively. For simplicity, hereafter, we consider the case $\boldsymbol{L}^{\rm OLS} = (\boldsymbol{X}^\top\boldsymbol{X})^{-1}\boldsymbol{X}^\top$.

[1] Prasad–Rao estimator. The Prasad–Rao estimator with $\boldsymbol{W} = \boldsymbol{I}_m$ and $\boldsymbol{L} = \boldsymbol{L}^{\rm OLS}$ is given by

$$
\widehat{\psi}^{\rm PR} = \max\left\{\frac{\boldsymbol{y}^\top\boldsymbol{P}\boldsymbol{y} - {\rm tr}\,(\boldsymbol{D}) + {\rm tr}\,\{(\boldsymbol{X}^\top\boldsymbol{X})^{-1}\boldsymbol{X}^\top\boldsymbol{D}\boldsymbol{X}\}}{m-p},\ 0\right\}, \quad (4.14)
$$

for $\boldsymbol{P} = \boldsymbol{I}_m - \boldsymbol{X}(\boldsymbol{X}^\top\boldsymbol{X})^{-1}\boldsymbol{X}^\top$. It can be confirmed that conditions (FH4), (FH5), (C3†) and (C4†) are satisfied under (FH1)-(FH3). Then, ${\rm Bias}(\widehat{\psi}^{\rm PR}) = o(m^{-1})$ and

$$\mathrm{Var}(\widehat{\psi}^{\mathrm{PR}}) = \frac{2}{m^2}\mathrm{tr}\,(\boldsymbol{\Sigma}^2) + \frac{\kappa_v m\psi^2 + \kappa_e \mathrm{tr}\,(\boldsymbol{D}^2)}{m^2}.$$

From Theorems 4.2 and 4.4, the second-order approximation of EBLUP $\widehat{\theta}_{\mathrm{PR}}^{\mathrm{EBLUP}}$ with $\widehat{\psi}^{\mathrm{PR}}$ is

$$\begin{aligned}\mathrm{MSE}(\psi, \widehat{\theta}_{\mathrm{PR}}^{\mathrm{EBLUP}}) &= g_1(\psi) + g_2(\psi)\\ &+ \frac{D_i^2}{m^2(\psi + D_i)^3}\left[2\mathrm{tr}\,(\boldsymbol{\Sigma}^2) + \kappa_e\{\mathrm{tr}\,(\boldsymbol{D}^2) + 2m\psi D_i\} - \kappa_v m\psi^2\right] + o(m^{-1}),\end{aligned}$$

for $g_1(\psi)$ and $g_2(\psi)$ given in (4.7). Also, from Theorem 4.3,

$$\mathrm{mse}(\widehat{\theta}_{\mathrm{PR}}^{\mathrm{EBLUP}}) = g_1(\widehat{\psi}) + g_2(\widehat{\psi}) + \frac{2D_i^2}{m^2(\widehat{\psi} + D_i)^3}\left[2\mathrm{tr}\,(\boldsymbol{\Sigma}^2) + \kappa_e\{\mathrm{tr}\,(\boldsymbol{D}^2) + m\widehat{\psi} D_i\}\right],$$

which does not depend on an estimator of κ_v. This implies that the second-order unbiased estimator $\mathrm{mse}(\widehat{\theta}^{\mathrm{EBLUP}})$ is robust against the distribution of v_i as shown by Lahiri and Rao (1994).

[2] OLS-based REML estimator. This estimator corresponds to $\boldsymbol{W} = \boldsymbol{\Sigma}^{-2}$ and $\boldsymbol{L} = \boldsymbol{L}^{\mathrm{OLS}}$, and the estimator is given by $\widehat{\psi}^{\mathrm{ORE}} = \max(\psi^*, 0)$, where ψ^* is the solution of the estimating equation

$$\boldsymbol{y}^\top \boldsymbol{P}\boldsymbol{\Sigma}^{-2}\boldsymbol{P}\boldsymbol{y} = \mathrm{tr}\,(\boldsymbol{P}\boldsymbol{\Sigma}^{-2}\boldsymbol{P}\boldsymbol{\Sigma}).$$

Since $\boldsymbol{W}_{(1)} = -2\boldsymbol{\Sigma}^{-3}$, from Theorem 4.1, it follows that

$$\begin{aligned}\mathrm{Var}(\widehat{\psi}^{\mathrm{ORE}}) &= \frac{1}{\{\mathrm{tr}\,(\boldsymbol{\Sigma}^{-2})\}^2}\{2\mathrm{tr}\,(\boldsymbol{\Sigma}^{-2}) + \kappa(\boldsymbol{\Sigma}^{-2}, \boldsymbol{\Sigma}^{-2})\} + o(m^{-1}),\\ \mathrm{Bias}(\widehat{\psi}^{\mathrm{ORE}}) &= \frac{2}{\{\mathrm{tr}\,(\boldsymbol{\Sigma}^{-2})\}^2}\left\{-\kappa(\boldsymbol{\Sigma}^{-3}, \boldsymbol{\Sigma}^{-2}) + \frac{\mathrm{tr}\,(\boldsymbol{\Sigma}^{-3})}{\mathrm{tr}\,(\boldsymbol{\Sigma}^{-2})}\kappa(\boldsymbol{\Sigma}^{-2}, \boldsymbol{\Sigma}^{-2})\right\} + o(m^{-1}),\end{aligned}$$

where $\kappa(\boldsymbol{\Sigma}^{-2}, \boldsymbol{\Sigma}^{-2}) = \kappa_e \mathrm{tr}\,(\boldsymbol{D}^2\boldsymbol{\Sigma}^{-4}) + \psi^2\kappa_v \mathrm{tr}\,(\boldsymbol{\Sigma}^{-4})$ and $\kappa(\boldsymbol{\Sigma}^{-3}, \boldsymbol{\Sigma}^{-2}) = \kappa_e \mathrm{tr}\,(\boldsymbol{D}^2\boldsymbol{\Sigma}^{-5}) + \psi^2\kappa_v \mathrm{tr}\,(\boldsymbol{\Sigma}^{-5})$. When $\kappa_v = \kappa_e = 0$, the estimator $\widehat{\psi}^{\mathrm{ORE}}$ is second-order unbiased, and we have $\mathrm{Var}(\widehat{\psi}^{\mathrm{ORE}}) = 2/\mathrm{tr}\,(\boldsymbol{\Sigma}^{-2}) + o(m^{-1})$. By the Cauchy–Schwarz inequality,

$$\frac{\mathrm{tr}\,(\boldsymbol{W}\boldsymbol{\Sigma}\boldsymbol{W}\boldsymbol{\Sigma})}{\{\mathrm{tr}\,(\boldsymbol{W})\}^2} \geq \frac{1}{\mathrm{tr}\,(\boldsymbol{\Sigma}^{-2})},$$

which implies, from Theorem 4.1, that $\widehat{\psi}^{\mathrm{ORE}}$ is asymptotically better than or equal to any estimators given by (4.3).

From Theorems 4.2 and 4.4, the second-order approximation of EBLUP $\widehat{\theta}_{\mathrm{ORE}}^{\mathrm{EBLUP}}$ with $\widehat{\psi}^{\mathrm{ORE}}$ is

$$\mathrm{MSE}(\psi, \widehat{\theta}_{\mathrm{ORE}}^{\mathrm{EBLUP}}) = g_1(\psi) + g_2(\psi) + \frac{D_i^2}{(\psi + D_i)^3\{\mathrm{tr}\,(\boldsymbol{\Sigma}^{-2})\}^2}\{2\mathrm{tr}\,(\boldsymbol{\Sigma}^{-2}) + \kappa(\boldsymbol{\Sigma}^{-2}, \boldsymbol{\Sigma}^{-2})\}$$
$$+ 2\frac{D_i^2\psi}{(\psi + D_i)^5\mathrm{tr}\,(\boldsymbol{\Sigma}^{-2})}(\kappa_e D_i - \kappa_v \psi) + o(m^{-1}).$$

From Theorem 4.3, the second-order unbiased estimator of the MSE is

$$\mathrm{mse}(\widehat{\theta}_{\mathrm{ORE}}^{\mathrm{EBLUP}}) = g_1(\widehat{\psi}) + g_2(\widehat{\psi}) + 2\frac{D_i^2}{(\widehat{\psi} + D_i)^3\{\mathrm{tr}\,(\widehat{\boldsymbol{\Sigma}}^{-1})\}^2}\{2\mathrm{tr}\,(\widehat{\boldsymbol{\Sigma}}^{-2}) + \kappa(\widehat{\boldsymbol{\Sigma}}^{-2}, \widehat{\boldsymbol{\Sigma}}^{-2}, \widehat{\kappa}_v)\}$$
$$+ 2\frac{D_i^2\widehat{\psi}}{(\widehat{\psi} + D_i)^5\mathrm{tr}\,(\widehat{\boldsymbol{\Sigma}}^{-2})}(\kappa_e D_i - \widehat{\kappa}_v\widehat{\psi}) - \frac{D_i^2}{(\widehat{\psi} + D_i)^2}\widehat{\mathrm{Bias}}(\widehat{\psi}^{\mathrm{ORE}}).$$

[3] OLS-based Fay–Herriot estimator. This estimator corresponds to $\boldsymbol{W} = \boldsymbol{\Sigma}^{-1}$ and $\boldsymbol{L} = \boldsymbol{L}^{\mathrm{OLS}}$, and the estimator is given by $\widehat{\psi}^{\mathrm{OFH}} = \max(\psi^*, 0)$, where ψ^* is the solution of the estimating equation

$$\boldsymbol{y}^\top \boldsymbol{P}\boldsymbol{\Sigma}^{-1}\boldsymbol{P}\boldsymbol{y} = \mathrm{tr}\,(\boldsymbol{P}\boldsymbol{\Sigma}^{-1}\boldsymbol{P}\boldsymbol{\Sigma}).$$

Since $\boldsymbol{W}_{(1)} = -\boldsymbol{\Sigma}^{-2}$, from Theorem 4.1, it follows that

$$\mathrm{Var}(\widehat{\psi}^{\mathrm{OFH}}) = \frac{1}{\{\mathrm{tr}\,(\boldsymbol{\Sigma}^{-1})\}^2}\{2m + \kappa(\boldsymbol{\Sigma}^{-1}, \boldsymbol{\Sigma}^{-1})\} + o(m^{-1}),$$
$$\mathrm{Bias}(\widehat{\psi}^{\mathrm{OFH}}) = -\frac{2\mathrm{tr}\,(\boldsymbol{\Sigma}^{-1}) + \kappa(\boldsymbol{\Sigma}^{-2}, \boldsymbol{\Sigma}^{-1})}{\{\mathrm{tr}\,(\boldsymbol{\Sigma}^{-1})\}^2} + \frac{\mathrm{tr}\,(\boldsymbol{\Sigma}^{-2})\{2m + \kappa(\boldsymbol{\Sigma}^{-1}, \boldsymbol{\Sigma}^{-1})\}}{\{\mathrm{tr}\,(\boldsymbol{\Sigma}^{-1})\}^3} + o(m^{-1}),$$

where $\kappa(\boldsymbol{\Sigma}^{-1}, \boldsymbol{\Sigma}^{-1}) = \kappa_e\mathrm{tr}\,(\boldsymbol{D}^2\boldsymbol{\Sigma}^{-2}) + \psi^2\kappa_v\mathrm{tr}\,(\boldsymbol{\Sigma}^{-2})$ and $\kappa(\boldsymbol{\Sigma}^{-2}, \boldsymbol{\Sigma}^{-1}) = \kappa_e\mathrm{tr}\,(\boldsymbol{D}^2\boldsymbol{\Sigma}^{-3}) + \psi^2\kappa_v\mathrm{tr}\,(\boldsymbol{\Sigma}^{-3})$. From Theorems 4.2 and 4.4, the second-order approximation of EBLUP $\widehat{\theta}_{\mathrm{OFH}}^{\mathrm{EBLUP}}$ with $\widehat{\psi}^{\mathrm{OFH}}$ is

$$\mathrm{MSE}(\psi, \widehat{\theta}_{\mathrm{OFH}}^{\mathrm{EBLUP}}) = g_1(\psi) + g_2(\psi) + \frac{D_i^2}{(\psi + D_i)^3\{\mathrm{tr}\,(\boldsymbol{\Sigma}^{-1})\}^2}\{2m + \kappa(\boldsymbol{\Sigma}^{-1}, \boldsymbol{\Sigma}^{-1})\}$$
$$+ 2\frac{D_i^2\psi}{(\psi + D_i)^4\mathrm{tr}\,(\boldsymbol{\Sigma}^{-1})}\left\{\kappa_e D_i - \kappa_v\psi\right\} + o(m^{-1}).$$

From Theorem 4.3, the second-order unbiased estimator of the MSE is

$$\mathrm{mse}(\widehat{\theta}_{\mathrm{OFH}}^{\mathrm{EBLUP}}) = g_1(\widehat{\psi}) + g_2(\widehat{\psi}) + 2\frac{D_i^2}{(\widehat{\psi} + D_i)^3\{\mathrm{tr}\,(\widehat{\boldsymbol{\Sigma}}^{-1})\}^2}\{2m + \kappa(\widehat{\boldsymbol{\Sigma}}^{-1}, \widehat{\boldsymbol{\Sigma}}^{-1}, \widehat{\kappa}_v)\}$$
$$+ 2\frac{D_i^2\widehat{\psi}}{(\widehat{\psi} + D_i)^4\mathrm{tr}\,(\widehat{\boldsymbol{\Sigma}}^{-1})}\left\{\kappa_e D_i - \widehat{\kappa}_v\widehat{\psi}\right\} - \frac{D_i^2}{(\widehat{\psi} + D_i)^2}\widehat{\mathrm{Bias}}(\widehat{\psi}^{\mathrm{OFH}}),$$

where $\kappa(\widehat{\boldsymbol{\Sigma}}^{-1}, \widehat{\boldsymbol{\Sigma}}^{-1}, \widehat{\kappa}_v)$ replaces κ_v with $\widehat{\kappa}_v$ in $\kappa(\widehat{\boldsymbol{\Sigma}}^{-1}, \widehat{\boldsymbol{\Sigma}}^{-1})$.

Proof of Theorem 4.2 The functions $g_3(\psi)$ and $g_4(\psi)$ are provided from Theorem 3.1. In fact, we have $\boldsymbol{d}_{(1)}^\top \boldsymbol{\Sigma} \boldsymbol{d}_{(1)} = D_i^{-1}(1-\gamma_i)^3$, which leads to the expression in (4.9). The function g_4 given in Theorem 3.1 is

$$\mathrm{E}[(\widehat{\psi}-\psi)\{D_i^{-1}(1-\gamma_i)^2 u_i + \boldsymbol{m}_1^\top \boldsymbol{u}\}\{\gamma_i u_i - v_i + (1-\gamma_i)\boldsymbol{x}_i^\top \boldsymbol{H}\boldsymbol{u}\}],$$

which is decomposed as

$$\begin{aligned} &D_i^{-1}(1-\gamma_i)^2 \mathrm{E}[(\widehat{\psi}-\psi)u_i(\gamma_i u_i - v_i)] + D_i^{-1}(1-\gamma_i)^3 \mathrm{E}[(\widehat{\psi}-\psi)u_i \boldsymbol{x}_i^\top \boldsymbol{H}\boldsymbol{u}] \\ &\quad + \mathrm{E}[(\widehat{\psi}-\psi)\boldsymbol{m}_1^\top \boldsymbol{u}(\gamma_i u_i - v_i)] + (1-\gamma_i)\mathrm{E}[(\widehat{\psi}-\psi)\boldsymbol{m}_1^\top \boldsymbol{u}\boldsymbol{x}_i^\top \boldsymbol{H}\boldsymbol{u}]. \end{aligned}$$

Since $\mathrm{E}[u_i(\gamma_i u_i - v_i)] = 0$, we have $\mathrm{E}[(\widehat{\psi}-\psi)u_i(\gamma_i u_i - v_i)] = \mathrm{E}[\widehat{\psi} u_i (\gamma_i u_i - v_i)]$, which leads to (4.10). The other terms can be seen to be of order $o(m^{-1})$. For example, it is demonstrated that $\mathrm{E}[(\widehat{\psi}-\psi)\boldsymbol{m}_1^\top \boldsymbol{u}(\gamma_i u_i - v_i)] = o(m^{-1})$ from condition (B4), because $\boldsymbol{m}_1^\top \boldsymbol{u}$ is written as $\sum_{j=1}^m c_j u_j$ for $c_j = O(m^{-1})$.

We now verify the conditions (B1)–(B3) are satisfied under (FH1)–(FH4). For condition (B1), from condition (FH4), it follows that

$$\begin{aligned} \mathrm{E}[(\boldsymbol{m}_1^\top \boldsymbol{u})^2(\widehat{\psi}-\psi)^2] &= \sum_{i,j}^m c_i c_j \mathrm{E}[u_i u_j (\widehat{\psi}-\psi)^2] \\ &= \sum_{i,j}^m c_i c_j \mathrm{E}[u_i u_j]\mathrm{E}[(\widehat{\psi}-\psi)^2] + \sum_{i,j}^m c_i c_j \mathrm{E}[u_i u_j R_0(u_i, u_j)] \\ &= \sum_{i=1}^m c_i^2 \mathrm{E}[u_i^2]\mathrm{E}[(\widehat{\psi}-\psi)^2] + o(m^{-1}), \end{aligned} \tag{4.15}$$

which satisfies condition (B1) provided $\mathrm{E}[(\widehat{\psi}-\psi)^2] = O(m^{-1})$.

For condition (B2), note that

$$\begin{aligned} &(\widehat{\boldsymbol{d}}-\boldsymbol{d})^\top \boldsymbol{u} - \boldsymbol{d}_{(1)}^\top \boldsymbol{u}(\widehat{\psi}-\psi) = (\widehat{\gamma}_i - \gamma_i)u_i \\ &\quad + \{(1-\widehat{\gamma}_i)\boldsymbol{x}_i^\top \widehat{\boldsymbol{H}} - (1-\gamma_i)\boldsymbol{x}_i^\top \boldsymbol{H}\}\boldsymbol{u} - \{D_i^{-1}(1-\gamma_i)^2 u_i + \boldsymbol{m}_1^\top \boldsymbol{u}\}(\widehat{\psi}-\psi), \end{aligned}$$

where $\widehat{\gamma}_i = \widehat{\psi}/(\widehat{\psi}+D_i)$ and $\widehat{\boldsymbol{H}} = (\boldsymbol{X}^\top \widehat{\boldsymbol{\Sigma}}^{-1}\boldsymbol{X})^{-1}\boldsymbol{X}^\top \widehat{\boldsymbol{\Sigma}}^{-1}$ for $\widehat{\boldsymbol{\Sigma}} = \widehat{\psi}\boldsymbol{I}_m + \boldsymbol{D}$. Thus,

$$\mathrm{E}[\{(\widehat{\boldsymbol{d}}-\boldsymbol{d})^\top \boldsymbol{u} - \boldsymbol{d}_{(1)}^\top \boldsymbol{u}(\widehat{\psi}-\psi)\}^2] \le 3\mathrm{E}[(\widehat{\gamma}_i-\gamma_i)^2 u_i^2] \tag{4.16}$$

$$+\, 3\mathrm{E}[\{(1-\widehat{\gamma}_i)\boldsymbol{x}_i^\top \widehat{\boldsymbol{H}}\boldsymbol{u} - (1-\gamma_i)\boldsymbol{x}_i^\top \boldsymbol{H}\boldsymbol{u}\}^2] \tag{4.17}$$

$$+\, 3\mathrm{E}[\{D_i^{-1}(1-\gamma_i)^2 u_i + \boldsymbol{m}_1^\top \boldsymbol{u}\}^2(\widehat{\psi}-\psi)^2].$$

The first term in RHS of (4.16) is evaluated as

$$\begin{aligned}\mathrm{E}[(\widehat{\gamma}_i - \gamma_i)^2 u_i^2] &= \mathrm{E}[(\widehat{\psi} + D_i)^{-2}(\psi + D_i)^{-2}(\widehat{\psi} - \psi)^2 u_i^2] \\ &\le D_i^{-2}(\psi + D_i)^{-2}\mathrm{E}[(\widehat{\psi} - \psi)^2 u_i^2],\end{aligned}$$

which is of order $o(m^{-1})$ from condition (FH4). For the third term in RHS of (4.16), from this result and (4.15), it follows that $\mathrm{E}[\{D_i^{-1}(1-\gamma_i)^2 u_i + \boldsymbol{m}_1^\top \boldsymbol{u}\}^2(\widehat{\psi} - \psi)^2] = o(m^{-1})$. For the second term in RHS of (4.16), note that

$$\begin{aligned}&(1-\widehat{\gamma}_i)\boldsymbol{x}_i^\top \widehat{\boldsymbol{H}}\boldsymbol{u} - (1-\gamma_i)\boldsymbol{x}_i^\top \boldsymbol{H}\boldsymbol{u} \\ &\quad = (\gamma_i - \widehat{\gamma}_i)\boldsymbol{x}_i^\top(\widehat{\boldsymbol{H}} - \boldsymbol{H})\boldsymbol{u} + (\gamma_i - \widehat{\gamma}_i)\boldsymbol{x}_i^\top \boldsymbol{H}\boldsymbol{u} + (1-\gamma_i)\boldsymbol{x}_i^\top(\widehat{\boldsymbol{H}} - \boldsymbol{H})\boldsymbol{u}.\end{aligned}$$

This implies that $\mathrm{E}[\{(1-\widehat{\gamma}_i)\boldsymbol{x}_i^\top \widehat{\boldsymbol{H}}\boldsymbol{u} - (1-\gamma_i)\boldsymbol{x}_i^\top \boldsymbol{H}\boldsymbol{u}\}^2] = o(m^{-1})$ provided $\mathrm{E}[(\widehat{\gamma}_i - \gamma_i)^2(\boldsymbol{x}_i^\top \boldsymbol{H}\boldsymbol{u})^2] = o(m^{-1})$, $\mathrm{E}[\{\boldsymbol{x}_i^\top(\widehat{\boldsymbol{H}} - \boldsymbol{H})\boldsymbol{u}\}^2] = o(m^{-1})$ and $\mathrm{E}[(\widehat{\gamma}_i - \gamma_i)^2\{\boldsymbol{x}_i^\top(\widehat{\boldsymbol{H}} - \boldsymbol{H})\boldsymbol{u}\}^2] = o(m^{-1})$. By the same arguments as in (4.15), we can show $\mathrm{E}[(\widehat{\gamma}_i - \gamma_i)^2(\boldsymbol{x}_i^\top \boldsymbol{H}\boldsymbol{u})^2] = o(m^{-1})$ from condition (FH4). Since $0 < \gamma_i < 1$ and $0 < \widehat{\gamma}_i < 1$, we have $\mathrm{E}[(\widehat{\gamma}_i - \gamma_i)^2\{\boldsymbol{x}_i^\top(\widehat{\boldsymbol{H}} - \boldsymbol{H})\boldsymbol{u}\}^2] < 4\mathrm{E}[\{\boldsymbol{x}_i^\top(\widehat{\boldsymbol{H}} - \boldsymbol{H})\boldsymbol{u}\}^2]$. Thus, it is sufficient to show that $\mathrm{E}[\{\boldsymbol{x}_i^\top(\widehat{\boldsymbol{H}} - \boldsymbol{H})\boldsymbol{u}\}^2] = o(m^{-1})$. To this end, it is noted that

$$\begin{aligned}(\boldsymbol{X}^\top\widehat{\boldsymbol{\Sigma}}^{-1}\boldsymbol{X})^{-1} &= (\boldsymbol{X}^\top\boldsymbol{\Sigma}^{-1}\boldsymbol{X})^{-1} - (\widehat{\psi} - \psi)(\boldsymbol{X}^\top\widehat{\boldsymbol{\Sigma}}^{-1}\boldsymbol{X})^{-1}(\boldsymbol{X}^\top\widehat{\boldsymbol{\Sigma}}^{-1}\boldsymbol{\Sigma}^{-1}\boldsymbol{X})(\boldsymbol{X}^\top\boldsymbol{\Sigma}^{-1}\boldsymbol{X})^{-1}, \\ \boldsymbol{X}^\top\widehat{\boldsymbol{\Sigma}}^{-1}\boldsymbol{u} &= \boldsymbol{X}^\top\boldsymbol{\Sigma}^{-1}\boldsymbol{u} - (\widehat{\psi} - \psi)\boldsymbol{X}^\top\widehat{\boldsymbol{\Sigma}}^{-1}\boldsymbol{\Sigma}^{-1}\boldsymbol{u} \\ &= \boldsymbol{X}^\top\boldsymbol{\Sigma}^{-1}\boldsymbol{u} - (\widehat{\psi} - \psi)\boldsymbol{X}^\top\boldsymbol{\Sigma}^{-2}\boldsymbol{u} + (\widehat{\psi} - \psi)^2\boldsymbol{X}^\top\widehat{\boldsymbol{\Sigma}}^{-1}\boldsymbol{\Sigma}^{-2}\boldsymbol{u}.\end{aligned}$$

Using these equations, we have

$$\begin{aligned}\boldsymbol{x}_i^\top(\widehat{\boldsymbol{H}} - \boldsymbol{H})\boldsymbol{u} = &-(\widehat{\psi} - \psi)\boldsymbol{x}_i^\top(\boldsymbol{X}^\top\boldsymbol{\Sigma}^{-1}\boldsymbol{X})^{-1}\boldsymbol{X}^\top\boldsymbol{\Sigma}^{-2}\boldsymbol{u} \\ &- (\widehat{\psi} - \psi)\boldsymbol{x}_i^\top(\boldsymbol{X}\widehat{\boldsymbol{\Sigma}}^{-1}\boldsymbol{X})^{-1}(\boldsymbol{X}^\top\widehat{\boldsymbol{\Sigma}}^{-1}\boldsymbol{\Sigma}^{-1}\boldsymbol{X})(\boldsymbol{X}^\top\boldsymbol{\Sigma}^{-1}\boldsymbol{X})^{-1}\boldsymbol{X}^\top\boldsymbol{\Sigma}^{-1}\boldsymbol{u} \\ &+ (\widehat{\psi} - \psi)^2\boldsymbol{x}_i^\top(\boldsymbol{X}^\top\widehat{\boldsymbol{\Sigma}}^{-1}\boldsymbol{X})^{-1}\boldsymbol{X}^\top\widehat{\boldsymbol{\Sigma}}^{-1}\boldsymbol{\Sigma}^{-2}\boldsymbol{u} \\ &+ (\widehat{\psi} - \psi)^2\boldsymbol{x}_i^\top(\boldsymbol{X}^\top\widehat{\boldsymbol{\Sigma}}^{-1}\boldsymbol{X})^{-1}(\boldsymbol{X}^\top\widehat{\boldsymbol{\Sigma}}^{-1}\boldsymbol{\Sigma}^{-1}\boldsymbol{X})(\boldsymbol{X}^\top\boldsymbol{\Sigma}^{-1}\boldsymbol{X})^{-1}\boldsymbol{X}^\top\boldsymbol{\Sigma}^{-2}\boldsymbol{u}.\end{aligned} \tag{4.18}$$

Hence, we can see that $\mathrm{E}[\{\boldsymbol{x}_i^\top(\widehat{\boldsymbol{H}} - \boldsymbol{H})\boldsymbol{u}\}^2] = o(m^{-1})$ if the expectations of squares of each term in (4.18) are of order $o(m^{-1})$. For the first term in (4.18), it is seen that $\mathrm{E}[(\widehat{\psi} - \psi)^2\{\boldsymbol{x}_i^\top(\boldsymbol{X}^\top\boldsymbol{\Sigma}^{-1}\boldsymbol{X})^{-1}\boldsymbol{X}^\top\boldsymbol{\Sigma}^{-2}\boldsymbol{u}\}^2] = o(m^{-1})$ from (4.15). By the Cauchy–Schwarz inequality,

$$\begin{aligned}&\{\boldsymbol{x}_i^\top(\boldsymbol{X}^\top\widehat{\boldsymbol{\Sigma}}^{-1}\boldsymbol{X})^{-1}(\boldsymbol{X}^\top\widehat{\boldsymbol{\Sigma}}^{-1}\boldsymbol{\Sigma}^{-1}\boldsymbol{X})(\boldsymbol{X}^\top\boldsymbol{\Sigma}^{-1}\boldsymbol{X})^{-1}\boldsymbol{X}^\top\boldsymbol{\Sigma}^{-1}\boldsymbol{u}\}^2 \\ &\le \boldsymbol{x}_i^\top(\boldsymbol{X}^\top\widehat{\boldsymbol{\Sigma}}^{-1}\boldsymbol{X})^{-1}(\boldsymbol{X}^\top\widehat{\boldsymbol{\Sigma}}^{-1}\boldsymbol{\Sigma}^{-1}\boldsymbol{X})^2(\boldsymbol{X}^\top\widehat{\boldsymbol{\Sigma}}^{-1}\boldsymbol{X})^{-1}\boldsymbol{x}_i \\ &\quad \times \boldsymbol{u}^\top\boldsymbol{\Sigma}^{-1}\boldsymbol{X}(\boldsymbol{X}^\top\boldsymbol{\Sigma}^{-1}\boldsymbol{X})^{-2}\boldsymbol{X}^\top\boldsymbol{\Sigma}^{-1}\boldsymbol{u} \\ &\le \frac{\boldsymbol{x}_i^\top\boldsymbol{x}_i}{D_L}\sum_{a,b}\frac{\boldsymbol{x}_a^\top(\boldsymbol{X}^\top\boldsymbol{\Sigma}^{-1}\boldsymbol{X})^{-2}\boldsymbol{x}_b}{(\psi + D_a)(\psi_b + D_b)}u_a u_b,\end{aligned}$$

because $X^\top\widehat{\Sigma}^{-1}\Sigma^{-1}X \le D_L^{-1}X^\top\widehat{\Sigma}^{-1}X$. Using the same arguments as in (4.15), we can see that the expectation of square of the second term is less than or equal to $D_L^{-1}x_i^\top x_i \mathrm{E}[(\widehat{\psi}-\psi)^2\sum_{a,b}C_{ab}u_a u_b]$, which is of order $o(m^{-1})$, where $C_{ab} = x_a^\top(X^\top\Sigma^{-1}X)^{-2}x_b/\{(\psi+D_a)(\psi+D_b)\}$. For the third term, by the Cauchy–Schwarz inequality,

$$
\begin{aligned}
&\{x_i^\top(X^\top\widehat{\Sigma}^{-1}X)^{-1}X^\top\widehat{\Sigma}^{-1}\Sigma^{-2}u\}^2] \\
&\le x_i^\top(X^\top\widehat{\Sigma}^{-1}X)^{-1}X^\top\widehat{\Sigma}^{-2}\Sigma^{-4}X(X^\top\widehat{\Sigma}^{-1}X)^{-1}x_i\cdot u^\top u.
\end{aligned}
$$

Note that

$$
\begin{aligned}
&\Big(X^\top\mathrm{diag}_j\Big(\frac{\widehat{\psi}+D_L}{\widehat{\psi}+D_j}\Big)X\Big)^{-1}X^\top\mathrm{diag}_j\Big(\frac{\widehat{\psi}+D_L}{(\widehat{\psi}+D_j)^2(\psi+D_j)^4}\Big)X \\
&\times\Big(X^\top\mathrm{diag}_j\Big(\frac{\widehat{\psi}+D_L}{\widehat{\psi}+D_j}\Big)X\Big)^{-1} \\
&\le\frac{1}{D_L^4}\Big(X^\top\mathrm{diag}_j\Big(\frac{\widehat{\psi}+D_L}{\widehat{\psi}+D_j}\Big)X\Big)^{-1}\le\frac{1}{D_L^4}\Big(X^\top\mathrm{diag}_j\Big(\frac{D_L}{D_j}\Big)X\Big)^{-1},
\end{aligned}
$$

where $\mathrm{diag}_j(a_j)=\mathrm{diag}(a_1,\ldots,a_m)$. Thus,

$$
\begin{aligned}
&\mathrm{E}[(\widehat{\psi}-\psi)^4\{x_i^\top(X^\top\widehat{\Sigma}^{-1}X)^{-1}X^\top\widehat{\Sigma}^{-1}\Sigma^{-2}u\}^2] \\
&\le\frac{1}{D_L^4}x_i^\top\Big(X^\top\mathrm{diag}_j\Big(\frac{D_L}{D_j}\Big)X\Big)^{-1}x_i\mathrm{E}[(\widehat{\psi}-\psi)^4u^\top u],
\end{aligned}
$$

which is of order $o(m^{-1})$ from condition (FH4). Similarly, we can show that

$$
\mathrm{E}[(\widehat{\psi}-\psi)^4\{x_i^\top(X^\top\widehat{\Sigma}^{-1}X)^{-1}(X^\top\widehat{\Sigma}^{-1}\Sigma^{-1}X)(X^\top\Sigma^{-1}X)^{-1}X^\top\Sigma^{-2}u\}^2]=o(m^{-1})
$$

from (FH4).

For condition (B3), noting that $d^\top u=\gamma_i u_i+(1-\gamma_i)x_i^\top Hu$ and $d_{(1)}^\top u = D_i^{-1}(1-\gamma_i)^2u_i - D_i^{-1}(1-\gamma_i)^2x_i^\top Hu + (1-ga_i)x_i^\top H_{(1)}u$, we have

$$
(\widehat{d}-d)^\top u - d_{(1)}^\top u(\widehat{\psi}-\psi) = (\widehat{\gamma}_i-\gamma_i)u_i - D_i^{-1}(1-\gamma_i)^2u_i(\widehat{\psi}-\psi)+d^*(u),
$$

where

$$
\begin{aligned}
d^*(u) &= (1-\widehat{\gamma}_i)x_i^\top\widehat{H}u-(1-\gamma_i)x_i^\top Hu + D_i^{-1}(1-\gamma_i)^2x_i^\top Hu(\widehat{\psi}-\psi) \\
&\quad -(1-\gamma_i)x_i^\top H_{(1)}u(\widehat{\psi}-\psi).
\end{aligned}
$$

It is observed that

$$
\begin{aligned}
&(\widehat{\gamma}_i - \gamma_i)u_i - D_i^{-1}(1-\gamma_i)^2 u_i(\widehat{\psi} - \psi) \\
&\quad = -D_i^{-2}(1-\gamma_i)^3(\widehat{\psi} - \psi)^2 u_i + D_i^{-3}(1-\gamma_i)^3(1-\widehat{\gamma}_i)(\widehat{\psi} - \psi)^3 u_i.
\end{aligned}
$$

Then, it is sufficient to show that $\mathrm{E}[(\widehat{\psi} - \psi)^2 u_i(\gamma_i u_i - v_i + (1-\gamma_i)\boldsymbol{x}_i^\top \boldsymbol{H}\boldsymbol{u})] = o(m^{-1})$, $\mathrm{E}[(1-\widehat{\gamma}_i)(\widehat{\psi} - \psi)^3 u_i(\gamma_i u_i - v_i + (1-\gamma_i)\boldsymbol{x}_i^\top \boldsymbol{H}\boldsymbol{u})] = o(m^{-1})$ and $\mathrm{E}[d(\boldsymbol{u})(\gamma_i u_i - v_i + (1-\gamma_i)\boldsymbol{x}_i^\top \boldsymbol{H}\boldsymbol{u})] = o(m^{-1})$. When c_js are scalars such that $c_j = O(m^{-1})$ and $\boldsymbol{x}_i^\top \boldsymbol{H}\boldsymbol{u} = \sum_{j=1}^m c_j u_j$, from (FH4), it follows that

$$
\begin{aligned}
&\mathrm{E}[(\widehat{\psi} - \psi)^2 u_i(\gamma_i u_i - v_i + (1-\gamma_i)\boldsymbol{x}_i^\top \boldsymbol{H}\boldsymbol{u})] \\
&\quad = \mathrm{E}[(\widehat{\psi} - \psi)^2]\mathrm{E}[u_i(\gamma_i u_i - v_i)] + \mathrm{E}[u_i(\gamma_i u_i - v_i)R_0(v_i, \varepsilon_i)] \\
&\quad + (1-\gamma_i)\sum_{j=1}^m \Big\{\mathrm{E}[(\widehat{\psi} - \psi)^2]c_j \mathrm{E}[u_i u_j] + c_j \mathrm{E}[u_i u_j R_0(u_i, u_j)]\Big\},
\end{aligned}
$$

which is of order $o(m^{-1})$, because $\mathrm{E}[u_i(\gamma_i u_i - v_i)] = 0$. We next note that

$$
\begin{aligned}
&|\mathrm{E}[(1-\widehat{\gamma}_i)(\widehat{\psi} - \psi)^3 u_i(\gamma_i u_i - v_i + (1-\gamma_i)\boldsymbol{x}_i^\top \boldsymbol{H}\boldsymbol{u})]| \\
&\quad \le \{\mathrm{E}[(1-\widehat{\gamma}_i)^2(\widehat{\psi} - \psi)^2]\mathrm{E}[(\widehat{\psi} - \psi)^4 u_i(\gamma_i u_i - v_i + (1-\gamma_i)\boldsymbol{x}_i^\top \boldsymbol{H}\boldsymbol{u})]\}^{1/2}
\end{aligned}
$$

by the Cauchy–Schwarz inequality. The condition (B4) guarantees it is of order $o(m^{-1})$. Finally, $d^*(\boldsymbol{u})$ can be rewritten as

$$
\begin{aligned}
&d^*(\boldsymbol{u}) \\
&= -\frac{D_i(\widehat{\psi} - \psi)}{(\psi + D_i)^2}\boldsymbol{x}_i^\top(\widehat{\boldsymbol{H}} - \boldsymbol{H})\boldsymbol{u} + \frac{D_i(\widehat{\psi} - \psi)^2}{(\widehat{\psi} + D_i)(\psi + D_i)^2}\boldsymbol{x}_i^\top \widehat{\boldsymbol{H}}\boldsymbol{u} \\
&+ \frac{D_i}{\psi + D_i}\boldsymbol{x}_i^\top\{\widehat{\boldsymbol{H}} - \boldsymbol{H} - \boldsymbol{H}_{(1)}(\widehat{\psi} - \psi)\}\boldsymbol{u} \\
&= -\frac{D_i(\widehat{\psi} - \psi)^2}{(\psi + D_i)^2}\boldsymbol{x}_i^\top \boldsymbol{H}_{(1)}^*\boldsymbol{u} + \frac{D_i(\widehat{\psi} - \psi)^2}{(\widehat{\psi} + D_i)(\psi + D_i)^2}\boldsymbol{x}_i^\top \widehat{\boldsymbol{H}}\boldsymbol{u} + \frac{D_i(\widehat{\psi} - \psi)^2}{2(\psi + D_i)}\boldsymbol{x}_i^\top \boldsymbol{H}_{(1,1)}^{**}\boldsymbol{u},
\end{aligned}
$$

for $\boldsymbol{H}_{(1)}^* = \boldsymbol{H}_{(1)}(\psi^*)$ and $\boldsymbol{H}_{(1,1)}^{**} = \boldsymbol{H}_{(1,1)}(\psi^{**})$ where ψ^* and ψ^{**} are on the line segment between ψ and $\widehat{\psi}$. By the Cauchy–Schwarz inequality,

$$
\begin{aligned}
&\mathrm{E}[(\widehat{\psi} - \psi)^2 \boldsymbol{x}_i^\top \widehat{\boldsymbol{H}}\boldsymbol{u}(\gamma_i u_i - v_i + (1-\gamma_i)\boldsymbol{x}_i^\top \boldsymbol{H}\boldsymbol{u})] \\
&\quad \le \{\mathrm{E}[(\widehat{\psi} - \psi)^2(\boldsymbol{x}_i^\top \widehat{\boldsymbol{H}}\boldsymbol{u})^2]\mathrm{E}[(\widehat{\psi} - \psi)^2(\gamma_i u_i - v_i + (1-\gamma_i)\boldsymbol{x}_i^\top \boldsymbol{H}\boldsymbol{u})^2]\}^{1/2}.
\end{aligned}
$$

The same arguments as used around (4.18) show that $\mathrm{E}[(\widehat{\psi} - \psi)^2(\boldsymbol{x}_i^\top \widehat{\boldsymbol{H}}\boldsymbol{u})^2] = o(m^{-1})$ and $\mathrm{E}[(\widehat{\psi} - \psi)^2(\gamma_i u_i - v_i + (1-\gamma_i)\boldsymbol{x}_i^\top \boldsymbol{H}\boldsymbol{u})^2] = O(m^{-1})$. The other terms can be evaluated similarly, and it is shown that $\mathrm{E}[d(\boldsymbol{u})(\gamma_i u_i - v_i + (1-\gamma_i)\boldsymbol{x}_i^\top \boldsymbol{H}\boldsymbol{u})] = o(m^{-1})$. Therefore, condition (B3) is satisfied, and the proof is complete. □

4.2 Basic Unit-Level Models

4.2.1 Nested Error Regression Model

When unit-level data are available, we can use a model for more in-depth analysis. Let $y_{i1}, \ldots, y_{in_i}$ be a unit-level sample from the ith area for $i = 1, \ldots, m$, and let $\boldsymbol{x}_{i1}, \ldots, \boldsymbol{x}_{in_i}$ be fixed vectors of covariates with/without the intercept term. The nested error regression (NER) model given in Battese et al. (1988) is described as

$$y_{ij} = \boldsymbol{x}_{ij}^\top \boldsymbol{\beta} + v_i + \varepsilon_{ij}, \quad j = 1, \ldots, n_i, \quad i = 1, \ldots, m, \tag{4.19}$$

where v_i and ε_{ij} are random effects and error terms, respectively, and are mutually independently distributed as $\mathrm{E}[v_i] = \mathrm{E}[\varepsilon_{ij}] = 0$, $\mathrm{Var}(v_i) = \tau^2$, $\mathrm{Var}(\varepsilon_{ij}) = \sigma^2$, $\mathrm{E}[v_i^4] = (\kappa_v + 3)\tau^4$, and $\mathrm{E}[\varepsilon_{ij}^4] = (\kappa_e + 3)\sigma^4$, $\boldsymbol{\beta}$ is a p-dimensional vector of unknown regression coefficients, and τ^2, σ^2, κ_v, and κ_e are unknown parameters. It is noted that v_i is a random effect depending on the ith area and common to all the observations y_{ij}s in the same area. This induces correlations among y_{ij}s, that is, $\mathrm{Cov}(y_{ij}, y_{ij'}) = \tau^2$ for $j \neq j'$, noting that observations in the different areas are still independent. Hence, the variances σ^2 and τ^2 are called 'within' and 'between' components of variance, respectively.

Let $\boldsymbol{y}_i = (y_{i1}, \ldots, y_{i,n_i})^\top$, $\boldsymbol{X}_i = (\boldsymbol{x}_{i1}, \ldots, \boldsymbol{x}_{i,n_i})^\top$, and $\boldsymbol{\epsilon}_i = (\varepsilon_{i1}, \ldots, \varepsilon_{i,n_i})^\top$. For $\boldsymbol{j}_k = (1, \ldots, 1)^\top \in \mathbb{R}^k$, the NER model is written as

$$\boldsymbol{y}_i = \boldsymbol{X}_i \boldsymbol{\beta} + \boldsymbol{j}_{n_i} v_i + \boldsymbol{\epsilon}_i, \quad i = 1, \ldots, m,$$

and the model belongs to the linear mixed-effects model (2.1), where $\boldsymbol{y}$, $\boldsymbol{X}$, $\boldsymbol{Z}$, $\boldsymbol{v}$, and $\boldsymbol{\epsilon}$ correspond to $\boldsymbol{y} = (\boldsymbol{y}_1^\top, \ldots, \boldsymbol{y}_m^\top)^\top$, $\boldsymbol{X} = (\boldsymbol{X}_1^\top, \ldots, \boldsymbol{X}_m^\top)^\top$, $\boldsymbol{Z} = \mathrm{bloc\,diag}(\boldsymbol{j}_{n_1}, \ldots, \boldsymbol{j}_{n_m})$, $\boldsymbol{v} = (v_1, \ldots, v_m)^\top$, and $\boldsymbol{\epsilon} = (\boldsymbol{\epsilon}_1^\top, \ldots, \boldsymbol{\epsilon}_m^\top)^\top$. Let $\boldsymbol{\Sigma} = \mathrm{block\,diag}(\boldsymbol{\Sigma}_1, \ldots, \boldsymbol{\Sigma}_m)$, where $\boldsymbol{\Sigma}_i = \tau^2 \boldsymbol{J}_{n_i} + \sigma^2 \boldsymbol{I}_{n_i}$ for $\boldsymbol{J}_{n_i} = \boldsymbol{j}_{n_i} \boldsymbol{j}_{n_i}^\top$. Also, $\boldsymbol{R}_v$ and $\boldsymbol{R}_e$ correspond to $\tau^2 \boldsymbol{I}_m$ and $\sigma^2 \boldsymbol{I}_N$ for $N = \sum_{j=1}^m n_j$.

Consider to predict $\theta_i = \boldsymbol{c}_i^\top \boldsymbol{\beta} + v_i$ for a vector of constants $\boldsymbol{c}_i \in \mathbb{R}^p$. A natural choice of $\boldsymbol{c}_i$ is $\overline{\boldsymbol{x}}_i = n_i^{-1} \sum_{j=1}^{n_i} \boldsymbol{x}_{ij}$. Let $\boldsymbol{e}_i$ be the m-variate vector such that the i-th element is one and the others are zero. Note that

$$\boldsymbol{\Sigma}_i^{-1} = \frac{1}{\sigma^2}\Big(\boldsymbol{I}_{n_i} - \frac{\tau^2}{\sigma^2 + n_i \tau^2} \boldsymbol{J}_{n_i}\Big),$$

$$\boldsymbol{e}_i^\top \boldsymbol{Z} \boldsymbol{\Sigma}^{-1} \boldsymbol{y} = \frac{1}{\sigma^2}\Big(\boldsymbol{j}_{n_i}^\top - \frac{n_i \tau^2}{\sigma^2 + n_i \tau^2} \boldsymbol{j}_{n_i}^\top\Big) \boldsymbol{y}_i.$$

Then from (2.11), it follows that the BLUP of θ_i is given by

$$\widetilde{\theta}_i^{\rm BLUP}(\boldsymbol{\psi}) = \boldsymbol{c}_i^\top \widetilde{\boldsymbol{\beta}}(\boldsymbol{\psi}) + \tau^2 \boldsymbol{e}_i^\top \boldsymbol{Z}^\top \boldsymbol{\Sigma}^{-1}\{\boldsymbol{y} - \boldsymbol{X}\widetilde{\boldsymbol{\beta}}(\boldsymbol{\psi})\}$$
$$= \boldsymbol{c}_i^\top \widetilde{\boldsymbol{\beta}}(\boldsymbol{\psi}) + \frac{n_i\tau^2}{\sigma^2 + n_i\tau^2}\left\{\overline{y}_i - \overline{\boldsymbol{x}}_i^\top \widetilde{\boldsymbol{\beta}}(\boldsymbol{\psi})\right\}, \quad (4.20)$$

where $\boldsymbol{\psi} = (\tau^2, \sigma^2)^\top$, $\widetilde{\boldsymbol{\beta}} = \widetilde{\boldsymbol{\beta}}(\boldsymbol{\psi}) = (\boldsymbol{X}^\top \boldsymbol{\Sigma}^{-1}\boldsymbol{X})^{-1}\boldsymbol{X}^\top \boldsymbol{\Sigma}^{-1}\boldsymbol{y}$ and $\overline{y}_i = \sum_{j=1}^{n_i} y_{ij}$.

The NER model is typically used in the framework of a finite population model. We assume that the ith area includes N_i units in total, but only n_i units are sampled. For simplicity, we assume sampling mechanism is the simple random sampling, so we do not consider survey weights. For all the units, we assume the following super-population model:

$$Y_{ij} = \boldsymbol{x}_{ij}^\top \boldsymbol{\beta} + v_i + \varepsilon_{ij}, \quad j = 1, \ldots, N_i, \quad i = 1, \ldots, m,$$

where Y_{ij} is the characteristic for the jth unit in the ith area. Without loss of generality, we assume the first n_i characteristics $Y_{i1}, \ldots, Y_{in_i}$ are observed as $y_{i1}, \ldots, y_{in_i}$, and the rest of characteristics $Y_{i,n_i+1}, \ldots, Y_{iN_i}$ are unobserved. The objective is to estimate the true mean $\overline{Y}_i = N_i^{-1}\sum_{j=1}^{N_i}$ based on all the observations. Under the setting, the true area mean is defined as

$$\frac{1}{N_i}\sum_{j=1}^{N_i} Y_{ij} = \frac{1}{N_i}\sum_{j=1}^{N_i}(\boldsymbol{x}_{ij}^\top \boldsymbol{\beta} + v_i + \varepsilon_{ij}) = \overline{\boldsymbol{X}}_i^\top \boldsymbol{\beta} + v_i + \frac{1}{N_i}\sum_{j=1}^{N_i}\varepsilon_{ij},$$

where $\overline{\boldsymbol{X}}_i = N_i^{-1}\sum_{j=1}^{N_i}\boldsymbol{x}_{ij}$. In practice, the total number of units N_i is very large although the number of sampled units n_i is not large. Then, the final term could be negligible, thereby we can define the mean parameter θ_i as $\theta_i = \overline{\boldsymbol{X}}_i^\top \boldsymbol{\beta} + v_i$. Hence, we can estimate θ_i only if we know the true mean vector $\overline{\boldsymbol{X}}_i$ of auxiliary information, which is actually often the case in practice. Under the model (4.19) with fixed $\boldsymbol{\psi} = (\tau^2, \sigma^2)^\top$, the best linear predictor of v_i is given by

$$\widetilde{v}_i(\boldsymbol{\psi}) = \frac{n_i\tau^2}{\sigma^2 + n_i\tau^2}\{\overline{y}_i - \overline{\boldsymbol{x}}_i^\top \widetilde{\boldsymbol{\beta}}(\boldsymbol{\psi})\},$$

where $\overline{y}_i = n_i^{-1}\sum_{j=1}^{n_i} y_{ij}$, $\overline{\boldsymbol{x}}_i = n_i^{-1}\sum_{j=1}^{n_i}\boldsymbol{x}_{ij}$ and $\widetilde{\boldsymbol{\beta}}(\boldsymbol{\psi})$ is the GLS estimator of $\boldsymbol{\beta}$. Then the BLUP of θ_i is given by $\widehat{\theta}_i^{\rm BLUP}(\boldsymbol{\psi}) = \overline{\boldsymbol{X}}_i^\top \widetilde{\boldsymbol{\beta}}(\boldsymbol{\psi}) + \widetilde{v}_i(\boldsymbol{\psi})$. More general methods for predicting population means are discussed in Jiang and Lahiri (2006).

Since variance components τ^2 and σ^2 are unknown, we need to estimate them. Various methods for estimating the variance components have been suggested. When $\boldsymbol{\psi}$ is estimated by $\widehat{\boldsymbol{\psi}} = (\widehat{\tau}^2, \widehat{\sigma}^2)^\top$, the substituted one $\widetilde{\theta}_i^{\rm BLUP}(\widehat{\boldsymbol{\psi}})$, denoted by $\widehat{\theta}_i^{\rm EBLUP}$, is the EBLUP of θ_i. The asymptotic properties of $\widehat{\boldsymbol{\psi}}$ and $\widehat{\theta}_i^{\rm EBLUP}$ are provided in the following section.

4.2.2 Asymptotic Properties of EBLUP

Let $\boldsymbol{\psi} = (\psi_1, \psi_2)^\top$ for $\psi_1 = \tau^2$ and $\psi_2 = \sigma^2$. In the NER model given in (4.19), the covariance matrix of $\boldsymbol{y}_i$ is $\boldsymbol{\Sigma}_i = \psi_1 \boldsymbol{J}_{n_i} + \psi_2 \boldsymbol{I}_{n_i}$ for $\boldsymbol{J}_{n_i} = \boldsymbol{j}_{n_i}\boldsymbol{j}_{n_i}^\top$, and the covariance matrix of $\boldsymbol{y} = (\boldsymbol{y}_1^\top, \ldots, \boldsymbol{y}_m^\top)^\top$ is $\boldsymbol{\Sigma} = \text{blcok diag}(\boldsymbol{\Sigma}_1, \ldots, \boldsymbol{\Sigma}_m) = \psi_1 \boldsymbol{G} + \psi_2 \boldsymbol{I}_N$ for $\boldsymbol{G} = \text{blcok diag}(\boldsymbol{J}_{n_1}, \ldots, \boldsymbol{J}_{n_m})$. We here consider the estimators given in (2.20), namely

$$\boldsymbol{y}^\top(\boldsymbol{I} - \boldsymbol{X}\boldsymbol{L})^\top \boldsymbol{W}_a(\boldsymbol{I} - \boldsymbol{X}\boldsymbol{L})\boldsymbol{y} - \text{tr}\,\{(\boldsymbol{I} - \boldsymbol{X}\boldsymbol{L})^\top \boldsymbol{W}_a(\boldsymbol{I} - \boldsymbol{X}\boldsymbol{L})\boldsymbol{\Sigma}\} = 0, \quad a = 1, 2. \tag{4.21}$$

From Theorem 2.5, we can provide the second-order approximations of the covariance matrix and the bias under conditions (C1)–(C5). The covariance matrix of $\widehat{\boldsymbol{\psi}}$ is approximated as

$$\mathbf{Cov}\,(\widehat{\boldsymbol{\psi}}) = \boldsymbol{A}^{-1}\boldsymbol{B}\boldsymbol{A}^{-1} + o(m^{-1}), \tag{4.22}$$

where

$$\boldsymbol{A} = \begin{pmatrix} \text{tr}\,(\boldsymbol{W}_1\boldsymbol{G}),\ \text{tr}\,(\boldsymbol{W}_1) \\ \text{tr}\,(\boldsymbol{W}_2\boldsymbol{G}),\ \text{tr}\,(\boldsymbol{W}_2) \end{pmatrix},$$

$$\boldsymbol{B} = 2\begin{pmatrix} \text{tr}\,(\boldsymbol{W}_1\boldsymbol{\Sigma}\boldsymbol{W}_1\boldsymbol{\Sigma}),\ \text{tr}\,(\boldsymbol{W}_1\boldsymbol{\Sigma}\boldsymbol{W}_2\boldsymbol{\Sigma}) \\ \text{tr}\,(\boldsymbol{W}_1\boldsymbol{\Sigma}\boldsymbol{W}_2\boldsymbol{\Sigma}),\ \text{tr}\,(\boldsymbol{W}_2\boldsymbol{\Sigma}\boldsymbol{W}_2\boldsymbol{\Sigma}) \end{pmatrix} + \begin{pmatrix} \kappa(\boldsymbol{W}_1, \boldsymbol{W}_1),\ \kappa(\boldsymbol{W}_1, \boldsymbol{W}_2) \\ \kappa(\boldsymbol{W}_2, \boldsymbol{W}_1),\ \kappa(\boldsymbol{W}_2, \boldsymbol{W}_2) \end{pmatrix},$$

for $\kappa(\boldsymbol{W}_a, \boldsymbol{W}_b) = \psi_1^2\kappa_v \sum_{j=1}^m (\boldsymbol{Z}^\top\boldsymbol{W}_a\boldsymbol{Z})_{jj}(\boldsymbol{Z}^\top\boldsymbol{W}_b\boldsymbol{Z})_{jj} + \psi_2^2\kappa_e \sum_{j=1}^N (\boldsymbol{W}_a)_{jj} (\boldsymbol{W}_b)_{jj}$. The second-order bias of $\widehat{\boldsymbol{\psi}}$ is approximated as

$$\text{E}[\widehat{\boldsymbol{\psi}} - \boldsymbol{\psi}] = \boldsymbol{A}^{-1}\left\{ \begin{pmatrix} \text{tr}\,(\boldsymbol{F}_1\boldsymbol{A}^{-1}) \\ \text{tr}\,(\boldsymbol{F}_2\boldsymbol{A}^{-1}) \end{pmatrix} + \frac{1}{2}\begin{pmatrix} \text{tr}\,(\boldsymbol{E}_1\boldsymbol{A}^{-1}\boldsymbol{B}\boldsymbol{A}^{-1}) \\ \text{tr}\,(\boldsymbol{E}_2\boldsymbol{A}^{-1}\boldsymbol{B}\boldsymbol{A}^{-1}) \end{pmatrix} \right\} + o(N^{-1}), \tag{4.23}$$

where

$$\boldsymbol{E}_a = -\begin{pmatrix} 2\text{tr}\,(\boldsymbol{W}_{a(1)}\boldsymbol{G}), & \text{tr}\,(\boldsymbol{W}_{a(1)}) + \text{tr}\,(\boldsymbol{W}_{a(2)}\boldsymbol{G}) \\ \text{tr}\,(\boldsymbol{W}_{a(2)}\boldsymbol{G}) + \text{tr}\,(\boldsymbol{W}_{a(1)}), & 2\text{tr}\,(\boldsymbol{W}_{a(2)}) \end{pmatrix},$$

$$\boldsymbol{F}_a = \begin{pmatrix} 2\text{tr}\,(\boldsymbol{W}_{a(1)}\boldsymbol{\Sigma}\boldsymbol{W}_1\boldsymbol{\Sigma}) + \kappa(\boldsymbol{W}_{a(1)}, \boldsymbol{W}_1),\ 2\text{tr}\,(\boldsymbol{W}_{a(1)}\boldsymbol{\Sigma}\boldsymbol{W}_2\boldsymbol{\Sigma}) + \kappa(\boldsymbol{W}_{a(1)}, \boldsymbol{W}_2) \\ 2\text{tr}\,(\boldsymbol{W}_{a(2)}\boldsymbol{\Sigma}\boldsymbol{W}_1\boldsymbol{\Sigma}) + \kappa(\boldsymbol{W}_{a(2)}, \boldsymbol{W}_1),\ 2\text{tr}\,(\boldsymbol{W}_{a(2)}\boldsymbol{\Sigma}\boldsymbol{W}_2\boldsymbol{\Sigma}) + \kappa(\boldsymbol{W}_{a(2)}, \boldsymbol{W}_2) \end{pmatrix}.$$

The BLUP of $\theta_i = \boldsymbol{c}_i^\top\boldsymbol{\beta} + v_i$ is given from (4.20) as

$$\widetilde{\theta}_i^{\text{BLUP}}(\boldsymbol{\psi}) = \boldsymbol{c}_i^\top\widetilde{\boldsymbol{\beta}}(\boldsymbol{\psi}) + \gamma_i\{\overline{y}_i - \overline{\boldsymbol{x}}_i^\top\widetilde{\boldsymbol{\beta}}(\boldsymbol{\psi})\}, \quad \gamma_i = \frac{n_i\psi_1}{\psi_2 + n_i\psi_1}.$$

Noting that

$$\boldsymbol{\Sigma}^{-1} = \frac{1}{\psi_2}\boldsymbol{I}_N - \frac{1}{\psi_2}\text{block diag}\Big(\frac{\psi_1}{\psi_2 + n_1\psi_1}\boldsymbol{J}_{n_1}, \ldots, \frac{\psi_1}{\psi_2 + n_m\psi_1}\boldsymbol{J}_{n_m}\Big),$$

we have $\boldsymbol{\Sigma}^{-1}\boldsymbol{Z}\boldsymbol{e}_i = \psi_2^{-1}\boldsymbol{j}_{n_i} - \psi_2^{-1}\gamma_i\boldsymbol{j}_{n_i} = (n_i\psi_1)^{-1}\gamma_i\boldsymbol{j}_{n_i}$. Then the notations $\boldsymbol{H}$ and $\boldsymbol{d}^\top\boldsymbol{y}$ in (3.2) correspond to $\boldsymbol{H} = (\boldsymbol{X}^\top\boldsymbol{\Sigma}^{-1}\boldsymbol{X})^{-1}\boldsymbol{X}^\top\boldsymbol{\Sigma}^{-1}$ and

$$\boldsymbol{d}^\top\boldsymbol{y} = n_i^{-1}\gamma_i\boldsymbol{j}_{n_i}^\top\boldsymbol{y}_i + (\boldsymbol{c}_i - \gamma_i\overline{\boldsymbol{x}}_i)^\top\boldsymbol{H}\boldsymbol{y} = \gamma_i\overline{y}_i + (\boldsymbol{c}_i - \gamma_i\overline{\boldsymbol{x}}_i)^\top\boldsymbol{H}\boldsymbol{y}.$$

From (3.5),

$$\begin{aligned} g_1(\boldsymbol{\psi}) &= \frac{\psi_1\psi_2}{\psi_2 + n_i\psi_1} = \frac{\psi_2}{n_i}\gamma_i, \\ g_2(\boldsymbol{\psi}) &= (\boldsymbol{c}_i - \gamma_i\boldsymbol{x}_i)^\top(\boldsymbol{X}^\top\boldsymbol{\Sigma}^{-1}\boldsymbol{X})^{-1}(\boldsymbol{c}_i - \gamma_i\boldsymbol{x}_i). \end{aligned} \tag{4.24}$$

For estimator $\widehat{\boldsymbol{\psi}}$ satisfying conditions (B1)–(B3), from Theorem 3.1, the MSE of $\widehat{\theta}_i^{EBLUP}$ is approximated as

$$\text{MSE}(\boldsymbol{\psi}, \widehat{\theta}_i^{\text{EBLUP}}) = g_1(\boldsymbol{\psi}) + g_2(\boldsymbol{\psi}) + g_3(\boldsymbol{\psi}) + 2g_4(\boldsymbol{\psi}) + o(N^{-1}), \tag{4.25}$$

where $g_3(\boldsymbol{\psi})$ and $g_4(\boldsymbol{\psi})$ in (3.8) and (3.9) can be expressed as

$$\begin{aligned} g_3(\boldsymbol{\psi}) &= \frac{n_i}{(\psi_2 + n_i\psi_1)^3}(\psi_2, -\psi_1)\mathbf{Cov}\,(\widehat{\boldsymbol{\psi}})\begin{pmatrix}\psi_2\\ -\psi_1\end{pmatrix} + o(m^{-1}), \\ g_4(\boldsymbol{\psi}) &= \gamma_i(1-\gamma_i)E\Big[\Big(\frac{\widehat{\psi}_1}{\psi_1} - \frac{\widehat{\psi}_2}{\psi_2}\Big)\overline{u}_i(\gamma_i\overline{u}_i - v_i)\Big], \end{aligned}$$

where $\overline{u}_i = n_i^{-1}\sum_{j=1}^{n_i} u_{ij}$ for $u_{ij} = y_{ij} - \boldsymbol{x}_{ij}^\top\boldsymbol{\beta} = v_i + \varepsilon_{ij}$. For the expression of $g_4(\boldsymbol{\psi})$, we used the equation $\mathrm{E}[\overline{u}_i(\gamma_i\overline{u}_i - v_i)] = 0$.

Under the conditions given in Theorem 3.4, a second-order unbiased estimator of the MSE of EBLUP is

$$\text{mse}(\widehat{\theta}^{\text{EBLUP}}) = g_1(\widehat{\boldsymbol{\psi}}) + g_2(\widehat{\boldsymbol{\psi}}) + 2g_3(\widehat{\boldsymbol{\psi}}) + 2g_4(\widehat{\boldsymbol{\psi}}) - g_{12}(\widehat{\boldsymbol{\psi}}), \tag{4.26}$$

where

$$\begin{aligned} g_{12}(\boldsymbol{\psi}) &= \frac{1}{(\psi_2 + \boldsymbol{n}_i\psi_1)^2}(\psi_2^2, n_i\psi_1^2)\mathrm{E}[\widehat{\boldsymbol{\psi}} - \boldsymbol{\psi}], \\ g_{13}(\boldsymbol{\psi}) &= -2g_3(\boldsymbol{\psi}). \end{aligned}$$

For the estimator $\widehat{\boldsymbol{\psi}}$ defined in (4.21), if the conditions (C1)–(C5) and (B1)-(B4) are satisfied, the function $g_3(\boldsymbol{\psi})$ is approximated as

$$g_3(\boldsymbol{\psi}) = \frac{n_i}{(\psi_2 + n_i\psi_1)^3}(\psi_2, -\psi_1)\boldsymbol{A}^{-1}\boldsymbol{B}\boldsymbol{A}^{-1}\begin{pmatrix}\psi_2\\ -\psi_1\end{pmatrix} + o(m^{-1}).$$

For $g_4(\boldsymbol{\psi})$, consider the case of $\boldsymbol{W}_k = \text{block diag}(\boldsymbol{W}_k^{(1)}, \ldots, \boldsymbol{W}_k^{(m)})$ for $k = 1, 2$, where $\boldsymbol{W}_k^{(i)}$ is an $n_i \times n_i$ matrix. Note that $\psi_2 \boldsymbol{\Sigma}_i^{-1} = \boldsymbol{I}_{n_i} - n_i^{-1}\gamma_i \boldsymbol{J}_{n_i}$, $\psi_2 \boldsymbol{\Sigma}_i^{-1} \boldsymbol{j}_{n_i} = (1-\gamma_i)\boldsymbol{j}_{n_i}$ and $n_i \psi_1 \boldsymbol{\Sigma}_i^{-1} \boldsymbol{j}_{n_i} = \gamma_i \boldsymbol{j}_{n_i}$. Then, $g_4(\boldsymbol{\psi})$ can be approximated as

$$g_4(\boldsymbol{\psi}) = \kappa_e \frac{\psi_1 \psi_2^2}{(\psi_2 + n_i \psi_1)^3} (\psi_2, -\psi_1) \boldsymbol{A}^{-1} \begin{pmatrix} \operatorname{tr}(\boldsymbol{W}_1^{(i)}) \\ \operatorname{tr}(\boldsymbol{W}_2^{(i)}) \end{pmatrix}$$
$$- \kappa_v \frac{n_i \psi_1^2 \psi_2}{(\psi_2 + n_i \psi_1)^3} (\psi_2, -\psi_1) \boldsymbol{A}^{-1} \begin{pmatrix} \operatorname{tr}(\boldsymbol{W}_1^{(i)} \boldsymbol{J}_{n_i}) \\ \operatorname{tr}(\boldsymbol{W}_2^{(i)} \boldsymbol{J}_{n_i}) \end{pmatrix}.$$

We treat the three specific estimators of $\boldsymbol{\psi}$, namely, the PR-type estimator, the FH-type estimator, the REML estimator which correspond to $(\boldsymbol{W}_1^{\mathrm{Q}}, \boldsymbol{W}_2^{\mathrm{Q}}) = (\boldsymbol{G}, \boldsymbol{I}_m)$, $(\boldsymbol{W}_1^{\mathrm{FH}}, \boldsymbol{W}_2^{\mathrm{FH}}) = ((\boldsymbol{\Sigma}^{-1}\boldsymbol{G} + \boldsymbol{G}\boldsymbol{\Sigma}^{-1})/2, \boldsymbol{\Sigma}^{-1})$, $(\boldsymbol{W}_1^{\mathrm{RE}}, \boldsymbol{W}_2^{\mathrm{RE}}) = (\boldsymbol{\Sigma}^{-1}\boldsymbol{G}\boldsymbol{\Sigma}^{-1}, \boldsymbol{\Sigma}^{-2})$, respectively. For simplicity, hereafter, we consider the case of $\boldsymbol{L}^{\mathrm{OLS}} = (\boldsymbol{X}^\top \boldsymbol{X})^{-1} \boldsymbol{X}^\top$, that is, the estimating equations in (4.21) are

$$\boldsymbol{y}^\top \boldsymbol{P} \boldsymbol{W}_a \boldsymbol{P} \boldsymbol{y} - \operatorname{tr}(\boldsymbol{P} \boldsymbol{W}_a \boldsymbol{P} \boldsymbol{\Sigma}) = 0, \quad \text{for } a = 1, 2,$$

for $\boldsymbol{P} = \boldsymbol{I}_N - \boldsymbol{X}(\boldsymbol{X}^\top \boldsymbol{X})^{-1} \boldsymbol{X}^\top$.

[1] PR-type estimator. The PR-type estimator $\widehat{\boldsymbol{\psi}}^{\mathrm{Q}} = (\widehat{\psi}_1^{\mathrm{Q}}, \widehat{\psi}_2^{\mathrm{Q}})^\top$ corresponds to $(\boldsymbol{W}_1^{\mathrm{Q}}, \boldsymbol{W}_2^{\mathrm{Q}}) = (\boldsymbol{G}, \boldsymbol{I}_m)$ for $\boldsymbol{L} = \boldsymbol{L}^{\mathrm{OLS}}$. This estimator is second-order unbiased and has the asymptotic covariance matrix $\mathbf{Cov}(\widehat{\boldsymbol{\psi}}^{\mathrm{Q}}) \approx \boldsymbol{A}_{\mathrm{Q}}^{-1} \boldsymbol{B}_{\mathrm{Q}} \boldsymbol{A}_{\mathrm{Q}}^{-1}$, where

$$\boldsymbol{A}_{\mathrm{Q}} = \begin{pmatrix} \operatorname{tr}(\boldsymbol{G}^2) & \operatorname{tr}(\boldsymbol{G}) \\ \operatorname{tr}(\boldsymbol{G}) & \operatorname{tr}(\boldsymbol{I}_N) \end{pmatrix} = \begin{pmatrix} \sum_j n_j^2 & N \\ N & N \end{pmatrix},$$

$$\boldsymbol{B}_{\mathrm{Q}} = 2 \begin{pmatrix} \operatorname{tr}(\boldsymbol{\Sigma}^2 \boldsymbol{G}^2) & \operatorname{tr}(\boldsymbol{\Sigma}^2 \boldsymbol{G}) \\ \operatorname{tr}(\boldsymbol{\Sigma}^2 \boldsymbol{G}) & \operatorname{tr}(\boldsymbol{\Sigma}^2) \end{pmatrix} + \begin{pmatrix} \kappa(\boldsymbol{G}, \boldsymbol{G}), & \kappa(\boldsymbol{G}, \boldsymbol{I}_N) \\ \kappa(\boldsymbol{G}, \boldsymbol{I}_N), & \kappa(\boldsymbol{I}_N, \boldsymbol{I}_N) \end{pmatrix}$$
$$= 2 \begin{pmatrix} \sum_j n_j^4 \psi_1^2/\gamma_j^2 & \sum_j n_j^3 \psi_1^2/\gamma_j^2 \\ \sum_j n_j^3 \psi_1^2/\gamma_j^2 & (N-m)\psi_2^2 + \sum_j n_j^2 \psi_1^2/\gamma_j^2 \end{pmatrix} + \psi_1^2 \kappa_v \begin{pmatrix} \sum_j n_j^4 & \sum_j n_j^3 \\ \sum_j n_j^3 & \sum_j n_j^2 \end{pmatrix} + \psi_2^2 \kappa_e N \boldsymbol{J}_2,$$

where $\sum_j$ denotes $\sum_{j=1}^m$. For the MSE of EBLUP of $\theta_i = \boldsymbol{c}_i^\top \boldsymbol{\beta} + v_i$, the second-order approximation and the second-order unbiased estimator are given in (4.25) and (4.26), where $g_{12}(\boldsymbol{\psi}) = o(m^{-1})$,

$$g_3(\boldsymbol{\psi}) = \frac{n_i}{(\psi_2 + n_i \psi_1)^3} (\psi_2, -\psi_1) \boldsymbol{A}_{\mathrm{Q}}^{-1} \boldsymbol{B}_{\mathrm{Q}} \boldsymbol{A}_{\mathrm{Q}}^{-1} \begin{pmatrix} \psi_2 \\ -\psi_1 \end{pmatrix} + o(m^{-1}),$$

$$g_4(\boldsymbol{\psi}) = \frac{n_i \psi_1 \psi_2}{(\psi_2 + n_i \psi_1)^3} \left\{ \kappa_e \psi_2 (\psi_2, -\psi_1) \boldsymbol{A}_{\mathrm{Q}}^{-1} \begin{pmatrix} 1 \\ 1 \end{pmatrix} - \kappa_v n_i \psi_1 (\psi_2, -\psi_1) \boldsymbol{A}_{\mathrm{Q}}^{-1} \begin{pmatrix} n_i \\ 1 \end{pmatrix} \right\}.$$

[2] Prasad–Rao estimator. As estimation methods other than ML and REML, Henderson's methods and Rao's MINQUE methods are well-known procedures in estimation of variance components. Especially, Henderson's methods provide explicit expressions of unbiased estimators. Prasad and Rao (1990) derived estimators with explicit forms using the Henderson method (III), which is given as follows: Let $\boldsymbol{P} = \boldsymbol{I}_N - \boldsymbol{X}(\boldsymbol{X}^\top \boldsymbol{X})^{-1}\boldsymbol{X}^\top$ and $\boldsymbol{P}_K = \boldsymbol{K} - \boldsymbol{K}\boldsymbol{X}(\boldsymbol{X}^\top \boldsymbol{K}\boldsymbol{X})^{-1}\boldsymbol{X}^\top \boldsymbol{K}$, where $\boldsymbol{K} = \text{block diag}(\boldsymbol{K}_1, \ldots, \boldsymbol{K}_m)$ for $\boldsymbol{K}_j = \boldsymbol{I}_{n_j} - n_j^{-1}\boldsymbol{J}_{n_j}$. Let $S = \boldsymbol{y}^\top \boldsymbol{P}\boldsymbol{y}$ and $S_1 = \boldsymbol{y}^\top \boldsymbol{P}_K \boldsymbol{y}$. Then, the Prasad–Rao estimator, denoted by $\widehat{\boldsymbol{\psi}}^{\text{PR}} = (\widehat{\psi}_1^{\text{PR}}, \widehat{\psi}_2^{\text{PR}})^\top$, is written as

$$\widehat{\psi}_1^{\text{PR}} = \max\left\{0, \ \frac{S - (N-p)\widehat{\psi}_2^U}{N*}\right\} \quad \text{and} \quad \widehat{\psi}_2^{\text{PR}} = \frac{S_1}{N-m-p},$$

where $N_* = N - \text{tr}\,\{(\boldsymbol{X}^\top \boldsymbol{X})^{-1}\sum_{j=1}^m n_j^2 \overline{\boldsymbol{x}}_j \overline{\boldsymbol{x}}_j^\top\}$. Clearly, this estimator is a second-order unbiased estimator. Since it does not belong to the class in (4.21), we need to calculate the covariance matrix using Lemma 2.1. It is noted that $\widehat{\psi}_1^{\text{PR}} - \psi_1$ and $\widehat{\psi}_2^{\text{PR}} - \psi_2$ can be approximated as

$$\widehat{\psi}_1^{\text{PR}} - \psi_1 = \frac{1}{N}\{\boldsymbol{u}^\top \boldsymbol{u} - \text{tr}\,(\boldsymbol{\Sigma})\} - \frac{1}{N-m}\{\boldsymbol{u}^\top \boldsymbol{K}\boldsymbol{u} - \text{tr}\,(\boldsymbol{\Sigma}\boldsymbol{K})\} + o_p(N^{-1}),$$

$$\widehat{\psi}_2^{\text{PR}} - \psi_2 = \frac{1}{N-m}\{\boldsymbol{u}^\top \boldsymbol{K}\boldsymbol{u} - \text{tr}\,(\boldsymbol{\Sigma}\boldsymbol{K})\} + o_p(N^{-1}).$$

It can be observed that $\text{Var}(\boldsymbol{u}^\top \boldsymbol{u}) = 2\sum_j(\psi_2 + n_j\psi_1)^2 + 2(N-m)\psi_2^2 + \kappa_v\psi_1^2\sum_j n_j^2 + \kappa_e\psi_2^2 N$, $\text{Var}(\boldsymbol{u}^\top \boldsymbol{K}\boldsymbol{u}) = 2(N-m)\psi_2^2 + \kappa_e\psi_2^2\sum_j(n_j-1)^2/n_j$ and $\text{Cov}(\boldsymbol{u}^\top \boldsymbol{u}, \boldsymbol{u}^\top \boldsymbol{K}\boldsymbol{u}) = 2(N-m)\psi_2^2 + \kappa_e\psi_2^2(N-m)$. Then, the second-order approximation of the covariance matrix of $\widehat{\boldsymbol{\psi}}^{\text{PR}}$ is

$$\begin{aligned}\mathbf{Cov}\,(\widehat{\boldsymbol{\psi}}^{\text{PR}}) \approx & \frac{2}{N(N-m)}\begin{pmatrix}\sum_j(\psi_2 + n_j\psi_1)^2(N-m)/N + 2\psi_2^2 m^2/N, & -m \\ -m, & N\psi_2^2\end{pmatrix} \\ & + \frac{\kappa_v\psi_1^2}{N^2}\begin{pmatrix}\sum_j n_j^2, & 0 \\ 0, & 0\end{pmatrix} \\ & + \frac{\kappa_e\psi_2^2\sum_j(n_j-1)^2/n_j}{(N-m)^2}\begin{pmatrix}1, & -1 \\ -1, & 1\end{pmatrix} + \frac{\kappa_e\psi_2^2}{N}\begin{pmatrix}-1, & 1 \\ 1, & 0\end{pmatrix}.\end{aligned}$$

For the MSE of EBLUP of $\theta_i = \boldsymbol{c}_i^\top \boldsymbol{\beta} + v_i$, it can be observed that $\text{E}[\boldsymbol{u}^\top \boldsymbol{u}\boldsymbol{u}_i^\top \boldsymbol{J}_{n_i}\boldsymbol{u}_i] = 2n_i(\psi_2 + n_i\psi_1)^2 + Nn_i(\psi_1 + \psi_2)(\psi_2 + n_i\psi_1) + \kappa_e\psi_2^2 n_i + \kappa_v\psi_1^2 n_i^3$, $\text{E}[\boldsymbol{u}^\top \boldsymbol{u}\boldsymbol{j}_{n_i}^\top \boldsymbol{u}_i v_i] = Nn_i(\psi_1 + \psi_2)\psi_1 + 2n_i\psi_1(\psi_2 + n_i\psi_1) + \kappa_n\psi_1^2 n_i^2$, $\text{E}[\boldsymbol{u}^\top \boldsymbol{K}\boldsymbol{u}\boldsymbol{u}_i^\top \boldsymbol{J}_{n_i}\boldsymbol{u}_i] = (N-m)n_i\psi_2(\psi_2 + n_i\psi_1)$ and $\text{E}[\boldsymbol{u}^\top \boldsymbol{K}\boldsymbol{u}\boldsymbol{j}_{n_i}^\top \boldsymbol{u}_i v_i] = (N-m)n_i\psi_1\psi_2 + 3n_i^2\psi_1^2 + \kappa_v\psi_1^2 n_i^2$. Then, the second-order approximation and the second-order unbiased estimator are given in (4.25) and (4.26), where $g_{12}(\boldsymbol{\psi}) = o(m^{-1})$,

$$g_3(\boldsymbol{\psi}) = \frac{n_i}{(\psi_2 + n_i\psi_1)^3}(\psi_2, -\psi_1)\mathbf{Cov}\,(\widehat{\boldsymbol{\psi}}^{\mathrm{PR}})\begin{pmatrix}\psi_2\\ -\psi_1\end{pmatrix} + o(m^{-1}),$$

$$g_4(\boldsymbol{\psi}) = -\kappa_v\frac{n_i^2\psi_1^2\psi_2^2}{N(\psi_2 + n_i\psi_1)^3} + \kappa_e\frac{\psi_1\psi_2^2}{(\psi_2 + n_i\psi_1)^3}\left\{\frac{\psi_2 n_i}{N} - \frac{(n_i - 1)(\psi_1 + \psi_2)}{N - m}\right\}.$$

In the balanced case of $n_1 = \cdots = n_m$, the coefficients of κ_v in $g_3(\boldsymbol{\psi})$ and $g_4(\boldsymbol{\psi})$ are the same, which implies that the second-order unbiased estimator of MSE does not depend on an estimate of κ_v, namely, it is robust against the distribution of v_is.

Hereafter, we assume that $\kappa_e = \kappa_v = 0$ for simplicity.

[3] OLS-based REML estimator. The OLS-based REML estimator $\widehat{\boldsymbol{\psi}}^{\mathrm{ORM}} = (\widehat{\psi}_1^{\mathrm{ORM}}, \widehat{\psi}_2^{\mathrm{ORM}})^\top$ correspond to $(\boldsymbol{W}_1^{\mathrm{RE}}, \boldsymbol{W}_2^{\mathrm{RE}}) = (\boldsymbol{\Sigma}^{-1}\boldsymbol{G}\boldsymbol{\Sigma}^{-1}, \boldsymbol{\Sigma}^{-2})$ for $\widehat{\boldsymbol{\beta}} = \widehat{\boldsymbol{\beta}}^{\mathrm{O}}$. As seen from Datta and Lahiri (2000), this estimator is second-order unbiased and has the asymptotic covariance matrix $\mathbf{Cov}\,(\widehat{\boldsymbol{\psi}}^{\mathrm{ORM}}) \approx 2\boldsymbol{A}_{\mathrm{RM}}^{-1}$, where

$$\boldsymbol{A}_{\mathrm{RM}} = \begin{pmatrix}\mathrm{tr}\,\{(\boldsymbol{\Sigma}^{-1}\boldsymbol{G})^2\}, & \mathrm{tr}\,(\boldsymbol{\Sigma}^{-2}\boldsymbol{G})\\ \mathrm{tr}\,(\boldsymbol{\Sigma}^{-2}\boldsymbol{G}), & \mathrm{tr}\,(\boldsymbol{\Sigma}^{-2})\end{pmatrix} = \frac{1}{\psi_2^2}\begin{pmatrix}\sum_j n_j^2(1-\gamma_i)^2, & \sum_j n_j(1-\gamma_i)^2\\ \sum_j n_j(1-\gamma_i)^2, & \sum_j(1-\gamma_i)^2 + N - m\end{pmatrix}.$$

For the MSE of EBLUP of $\theta_i = \boldsymbol{c}_i^\top\boldsymbol{\beta} + v_i$, the second-order approximation and the second-order unbiased estimator are given in (4.25) and (4.26), where $g_{12}(\boldsymbol{\psi}) = o(m^{-1})$, $g_4(\boldsymbol{\psi}) = o(m^{-1})$ and

$$g_3(\boldsymbol{\psi}) = 2\frac{n_i}{(\psi_2 + n_i\psi_1)^3}(\psi_2, -\psi_1)\boldsymbol{A}_{\mathrm{RM}}^{-1}\begin{pmatrix}\psi_2\\ -\psi_1\end{pmatrix} + o(m^{-1}).$$

[4] OLS-based FH-type estimator. The OLS-based FH-type estimators $\widehat{\boldsymbol{\psi}}^{\mathrm{OFH}} = (\widehat{\psi}_1^{\mathrm{OFH}}, \widehat{\psi}_2^{\mathrm{OFH}})^\top$ correspond to $(\boldsymbol{W}_1^{\mathrm{FH}}, \boldsymbol{W}_2^{\mathrm{FH}}) = ((\boldsymbol{\Sigma}^{-1}\boldsymbol{G} + \boldsymbol{G}\boldsymbol{\Sigma}^{-1})/2, \boldsymbol{\Sigma}^{-1})$ for $\widehat{\boldsymbol{\beta}} = \widehat{\boldsymbol{\beta}}^{\mathrm{O}}$. The OLS-based FH estimator $\widehat{\boldsymbol{\psi}}^{\mathrm{OFH}}$ have the covariance matrix $\mathbf{Cov}\,(\widehat{\boldsymbol{\psi}}^{\mathrm{OFH}}) \approx \boldsymbol{A}_{\mathrm{FH}}^{-1}\boldsymbol{B}_{\mathrm{FH}}\boldsymbol{A}_{\mathrm{FH}}^{-1}$, where

$$\boldsymbol{A}_{\mathrm{FH}} = \begin{pmatrix}\mathrm{tr}\,(\boldsymbol{\Sigma}^{-1}\boldsymbol{G}^2) & \mathrm{tr}\,(\boldsymbol{\Sigma}^{-1}\boldsymbol{G})\\ \mathrm{tr}\,(\boldsymbol{\Sigma}^{-1}\boldsymbol{G}) & \mathrm{tr}\,(\boldsymbol{\Sigma}^{-1})\end{pmatrix} = \frac{1}{\psi_2}\begin{pmatrix}\sum_j n_j^2(1-\gamma_j) & \sum_j n_j(1-\gamma_j)\\ \sum_j n_j(1-\gamma_j) & \sum_j(1-\gamma_j) + N - m\end{pmatrix},$$

$$\boldsymbol{B}_{\mathrm{FH}} = 2\begin{pmatrix}\mathrm{tr}\,(\boldsymbol{\Sigma}^{-1}\boldsymbol{G}\boldsymbol{\Sigma}\boldsymbol{G} + \boldsymbol{G}^2)/2 & \mathrm{tr}\,(\boldsymbol{G})\\ \mathrm{tr}\,(\boldsymbol{G}) & \mathrm{tr}\,(\boldsymbol{I}_N)\end{pmatrix} = 2\begin{pmatrix}\sum_j n_j^2 & N\\ N & N\end{pmatrix}.$$

The second-order bias is

$$\mathrm{E}[\widehat{\boldsymbol{\psi}}^{\mathrm{OFH}} - \boldsymbol{\psi}] = \boldsymbol{A}_{\mathrm{FH}}^{-1}\left\{\begin{pmatrix}\mathrm{tr}\,(\boldsymbol{F}_1\boldsymbol{A}_{\mathrm{FH}}^{-1})\\ \mathrm{tr}\,(\boldsymbol{F}_2\boldsymbol{A}_{\mathrm{FH}}^{-1})\end{pmatrix} + \frac{1}{2}\begin{pmatrix}\mathrm{tr}\,(\boldsymbol{E}_1\boldsymbol{A}_{\mathrm{FH}}^{-1}\boldsymbol{B}_{\mathrm{FH}}\boldsymbol{A}_{\mathrm{FH}}^{-1})\\ \mathrm{tr}\,(\boldsymbol{E}_2\boldsymbol{A}_{\mathrm{FH}}^{-1}\boldsymbol{B}_{\mathrm{FH}}\boldsymbol{A}_{\mathrm{FH}}^{-1})\end{pmatrix}\right\} + o(N^{-1}),$$

where

$$
\boldsymbol{E}_1 = \frac{2}{\psi_2^2}\begin{pmatrix} \sum_j n_j^3(1-\gamma_j)^2, & \sum_j n_j^2(1-\gamma_j)^2 \\ \sum_j n_j^2(1-\gamma_j)^2, & \sum_j n_j(1-\gamma_j)^2 \end{pmatrix},
$$

$$
\boldsymbol{E}_2 = \frac{2}{\psi_2^2}\begin{pmatrix} \sum_j n_j^2(1-\gamma_j)^2, & \sum_j n_j(1-\gamma_j)^2 \\ \sum_j n_j(1-\gamma_j)^2, & \sum_j(1-\gamma_j)^2 + N - m \end{pmatrix},
$$

$$
\boldsymbol{F}_1 = -\frac{2}{\psi_2}\begin{pmatrix} \sum_j n_j^3(1-\gamma_j), & \sum_j n_j^2(1-\gamma_j) \\ \sum_j n_j^2(1-\gamma_j), & \sum_j n_j(1-\gamma_j) \end{pmatrix},
$$

$$
\boldsymbol{F}_2 = -\frac{2}{\psi_2}\begin{pmatrix} \sum_j n_j^2(1-\gamma_j), & \sum_j n_j(1-\gamma_j) \\ \sum_j n_j(1-\gamma_j), & \sum_j(1-\gamma_j) + N - m \end{pmatrix}.
$$

References

Battese GE, Harter RM, Fuller WA (1988) An error-components model for prediction of county crop areas using survey and satellite data. J. Am. Statist. Assoc. 83:28–36

Efron B, Morris C (1975) Data analysis using Stein's estimator and its generalizations. J. Am. Statist. Assoc. 70:311–319

Fay RE, Herriot R (1979) Estimates of income for small places: an application of James-Stein procedures to census data. J. Am. Statist. Assoc. 74:269–277

Henderson CR (1950) Estimation of genetic parameters. Ann. Math. Statist. 21:309–310

James W, Stein C (1961) Estimation with quadratic loss. In: Proceedings of the fourth Berkeley symposium on mathematical statistics and probability, vol 1. University of California Press, Berkeley. James and Stein, C, pp 361–379

Jiang J, Lahiri P (2006) Mixed model prediction and small area estimation (with discussions). Test 15:1–96

Lahiri P, Rao JNK (1995) Robust estimation of mean squared error of small area estimators. J. Am. Statist. Assoc. 90:758–766

Rao JNK, Molina I (2015) Small area estimation, 2nd edn. Wiley

Stein C (1956) Inadmissibility of the usual estimator for the mean of a multivariate normal distribution. In: Proceedings of the Third Berkeley symposium on mathematical statistics and probability, vol 1. University of California University, Berkeley, pp 197–206

Chapter 5
Hypothesis Tests and Variable Selection

It is important to find and select significant variables in regression in the linear mixed models. For example, the prediction error of the EBLUP in the Fay–Herriot model increases in the number of explanatory variables as seen from Theorems 4.2 and (4.7). In this chapter, we explain the two approaches to selecting significant variables: hypothesis tests and variable selection criteria.

5.1 Test Procedures for a Linear Hypothesis on Regression Coefficients

Consider the linear mixed models described in (2.1) with assuming the normality of $\boldsymbol{v}$ and $\boldsymbol{\epsilon}$. Then, the marginal distribution of $\boldsymbol{y}$ is the general linear regression model $\mathrm{N}(\boldsymbol{X}\boldsymbol{\beta}, \boldsymbol{\Sigma})$ where $\boldsymbol{\Sigma} = \boldsymbol{\Sigma}(\boldsymbol{\psi})$ for $\boldsymbol{\psi} = (\psi_1, \ldots, \psi_q)^\top$. Then we consider the problem of testing linear hypothesis given by

$$H_0 \ : \ \boldsymbol{R}\boldsymbol{\beta} = \boldsymbol{r},$$

where $\boldsymbol{R}$ is an $r \times p$ known matrix with rank r, $r \le p$, and $\boldsymbol{r}$ is an $r \times 1$ vector.

In general, the three test procedures are known for testing hypotheses: the Wald, Lagrange multiplier (LM), and likelihood ratio (LR) test statistics. These are defined as follows in the general framework of testing $H_0 : \boldsymbol{k}(\boldsymbol{\xi}) = \boldsymbol{0}$ against $H_1 : \boldsymbol{k}(\boldsymbol{\xi}) \neq \boldsymbol{0}$, where $\boldsymbol{\xi}$ is a k_0-dimensional unknown vector and $\boldsymbol{k}(\boldsymbol{\xi})$ is a function from $\mathbb{R}^{k_0}$ to $\mathbb{R}^{k_1}$ for $k_1 \le k_0$. When a random variable $\boldsymbol{X}$ has a likelihood function $L(\boldsymbol{\xi} \mid \boldsymbol{X})$ or the log-likelihood $\ell(\boldsymbol{\xi} \mid \boldsymbol{X})$, the unrestricted ML estimator $\widehat{\boldsymbol{\xi}}^{\mathrm{ML}}$ of $\boldsymbol{\xi}$ is the solution of the equation $\boldsymbol{s}(\boldsymbol{\xi}) = \boldsymbol{0}$ for the score function $\boldsymbol{s}(\boldsymbol{\xi}) = \partial\ell(\boldsymbol{\xi}|\boldsymbol{X})/\partial\boldsymbol{\xi}$. For the restricted

S. Sugasawa and T. Kubokawa, *Mixed-Effects Models and Small Area Estimation*, JSS Research Series in Statistics, https://doi.org/10.1007/978-981-19-9486-9_5

ML estimator under the hypothesis $H_0 : \boldsymbol{k}(\boldsymbol{\xi}) = \mathbf{0}$, we use the Lagrange multiplier $g(\boldsymbol{\xi}) = \ell(\boldsymbol{\xi} \mid \boldsymbol{X}) - \boldsymbol{\lambda}^\top \boldsymbol{k}(\boldsymbol{\xi})$. Since

$$\frac{\partial g(\boldsymbol{\xi})}{\partial \boldsymbol{\xi}} = \boldsymbol{s}(\boldsymbol{\xi}) - \{\boldsymbol{K}(\boldsymbol{\xi})\}^\top \boldsymbol{\lambda}(\boldsymbol{\xi})$$

for $\boldsymbol{K}(\boldsymbol{\xi}) = \partial \boldsymbol{k}(\boldsymbol{\xi})/\partial \boldsymbol{\xi}^\top$, the restricted ML estimator $\widehat{\boldsymbol{\xi}}_{\mathrm{R}}^{\mathrm{ML}}$ is the solution of the equations $\boldsymbol{s}(\boldsymbol{\xi}) - \{\boldsymbol{K}(\boldsymbol{\xi})\}^\top \boldsymbol{\lambda} = \mathbf{0}$ and $\boldsymbol{k}(\boldsymbol{\xi}) = \mathbf{0}$. Let $\widehat{\boldsymbol{\lambda}} = \widehat{\boldsymbol{\lambda}}\left(\widehat{\boldsymbol{\xi}}_{\mathrm{R}}^{\mathrm{ML}}\right)$ be the solution of $\boldsymbol{\lambda}$ derived from these equations. Then, the Wald, LM, and LR test statistics are given by

$$T_{\mathrm{W}} = \boldsymbol{k}\left(\widehat{\boldsymbol{\xi}}^{\mathrm{ML}}\right)^\top \left[\boldsymbol{K}\left(\widehat{\boldsymbol{\xi}}^{\mathrm{ML}}\right)\left\{\boldsymbol{I}\left(\widehat{\boldsymbol{\xi}}^{\mathrm{ML}}\right)\right\}^{-1} \boldsymbol{K}\left(\widehat{\boldsymbol{\xi}}^{\mathrm{ML}}\right)^\top\right]^{-1} \boldsymbol{k}\left(\widehat{\boldsymbol{\xi}}^{\mathrm{ML}}\right),$$

$$T_{\mathrm{LM}} = \boldsymbol{s}\left(\widehat{\boldsymbol{\xi}}_{\mathrm{R}}^{\mathrm{ML}}\right)^\top \left\{\boldsymbol{I}\left(\widehat{\boldsymbol{\xi}}_{\mathrm{R}}^{\mathrm{ML}}\right)\right\}^{-1} \boldsymbol{s}\left(\widehat{\boldsymbol{\xi}}_{\mathrm{R}}^{\mathrm{ML}}\right) = \widehat{\boldsymbol{\lambda}}^\top \boldsymbol{K}\left(\widehat{\boldsymbol{\xi}}_{\mathrm{R}}^{\mathrm{ML}}\right)\left\{\boldsymbol{I}\left(\widehat{\boldsymbol{\xi}}_{\mathrm{R}}^{\mathrm{ML}}\right)\right\}^{-1} \boldsymbol{K}\left(\widehat{\boldsymbol{\xi}}_{\mathrm{R}}^{\mathrm{ML}}\right)^\top \widehat{\boldsymbol{\lambda}},$$

$$T_{\mathrm{LR}} = -2\left\{\ell\left(\widehat{\boldsymbol{\xi}}_{\mathrm{R}}^{\mathrm{ML}} \mid \boldsymbol{X}\right) - \ell\left(\widehat{\boldsymbol{\xi}}^{\mathrm{ML}} \mid \boldsymbol{X}\right)\right\},$$

where $\boldsymbol{I}(\boldsymbol{\xi}) = \mathrm{E}[\boldsymbol{s}(\boldsymbol{\xi})\boldsymbol{s}(\boldsymbol{\xi})^\top]$ is the Fisher information matrix. The LM test is also called the score test or the Rao test. These test statistics converge to the chi-square distribution with k_1 degrees of freedom under H_0.

Return back to the problem of testing the linear hypothesis $H_0 : \boldsymbol{R\beta} = \boldsymbol{r}$ in the general linear regression model $\mathrm{N}(\boldsymbol{X\beta}, \boldsymbol{\Sigma})$. The log-likelihood is $\ell(\boldsymbol{\beta}, \boldsymbol{\psi} \mid \boldsymbol{y}) = -2^{-1} \log|\boldsymbol{\Sigma}(\boldsymbol{\psi})| - 2^{-1}(\boldsymbol{y} - \boldsymbol{X\beta})^\top \boldsymbol{\Sigma}(\boldsymbol{\psi})^{-1}(\boldsymbol{y} - \boldsymbol{X\beta})$. The unrestricted ML estimators $\widehat{\boldsymbol{\beta}}^{\mathrm{ML}}$ and $\widehat{\boldsymbol{\psi}}^{\mathrm{ML}}$ of $\boldsymbol{\beta}$ and $\boldsymbol{\psi}$ are the solutions of the equations $\partial \ell(\boldsymbol{\beta}, \boldsymbol{\psi} \mid \boldsymbol{y})/\partial \boldsymbol{\beta} = \mathbf{0}$ and $\partial \ell(\boldsymbol{\beta}, \boldsymbol{\psi} \mid \boldsymbol{y})/\partial \boldsymbol{\psi} = \mathbf{0}$, where $\widehat{\boldsymbol{\beta}}^{\mathrm{ML}} = \widetilde{\boldsymbol{\beta}}(\widehat{\boldsymbol{\psi}}^{\mathrm{ML}})$ for

$$\widetilde{\boldsymbol{\beta}}(\boldsymbol{\psi}) = (\boldsymbol{X}^\top \boldsymbol{\Sigma}(\boldsymbol{\psi})^{-1} \boldsymbol{X})^{-1} \boldsymbol{X}^\top \boldsymbol{\Sigma}(\boldsymbol{\psi})^{-1} \boldsymbol{y}.$$

On the other hand, the restricted ML estimators $\widehat{\boldsymbol{\beta}}_{\mathrm{R}}^{\mathrm{ML}}$ and $\widehat{\boldsymbol{\psi}}_{\mathrm{R}}^{\mathrm{ML}}$ under $\boldsymbol{R\beta} = \boldsymbol{r}$ are the solutions of the equations $\partial \ell(\boldsymbol{\beta}, \boldsymbol{\psi} \mid \boldsymbol{y})/\partial \boldsymbol{\beta} - \boldsymbol{R}^\top \boldsymbol{\lambda} = \mathbf{0}$, $\partial \ell(\boldsymbol{\beta}, \boldsymbol{\psi} \mid \boldsymbol{y})/\partial \boldsymbol{\psi} = \mathbf{0}$ and $\boldsymbol{R\beta} = \boldsymbol{r}$, where $\widehat{\boldsymbol{\beta}}_{\mathrm{R}}^{\mathrm{ML}} = \widetilde{\boldsymbol{\beta}}_{\mathrm{R}}(\widehat{\boldsymbol{\psi}}_{\mathrm{R}}^{\mathrm{ML}})$ for

$$\widetilde{\boldsymbol{\beta}}_{\mathrm{R}}(\boldsymbol{\psi}) = \widetilde{\boldsymbol{\beta}}(\boldsymbol{\psi}) - (\boldsymbol{X}^\top \boldsymbol{\Sigma}(\boldsymbol{\psi})^{-1} \boldsymbol{X})^{-1} \boldsymbol{R}^\top \boldsymbol{W}(\boldsymbol{\psi})(\boldsymbol{R}\widetilde{\boldsymbol{\beta}}(\boldsymbol{\psi}) - \boldsymbol{r}),$$

for $\boldsymbol{W}(\boldsymbol{\psi}) = [\boldsymbol{R}(\boldsymbol{X}^\top \boldsymbol{\Sigma}(\boldsymbol{\psi})^{-1} \boldsymbol{X})^{-1} \boldsymbol{R}^\top]^{-1}$. Under the hypothesis H_0, the statistic $\boldsymbol{R}\widehat{\boldsymbol{\beta}}^{\mathrm{ML}} - \boldsymbol{r}$ converges in distribution to $\mathrm{N}(\mathbf{0}, \boldsymbol{W}(\boldsymbol{\psi})^{-1})$, which provides the Wald test. Note that the score function is $\boldsymbol{X}^\top \boldsymbol{\Sigma}^{-1}(\boldsymbol{y} - \boldsymbol{X\beta})$ and the Fisher information is $\boldsymbol{I}(\boldsymbol{\beta}) = \boldsymbol{X}^\top \boldsymbol{\Sigma}^{-1} \boldsymbol{X}$. Since $\boldsymbol{X}^\top \boldsymbol{\Sigma}(\widehat{\boldsymbol{\psi}}_{\mathrm{R}}^{\mathrm{ML}})^{-1}(\boldsymbol{y} - \boldsymbol{X}\widetilde{\boldsymbol{\beta}}(\widehat{\boldsymbol{\psi}}_{\mathrm{R}}^{\mathrm{ML}})) = \mathbf{0}$, we have $\boldsymbol{X}^\top \boldsymbol{\Sigma}(\widehat{\boldsymbol{\psi}}_{\mathrm{R}}^{\mathrm{ML}})^{-1}(\boldsymbol{y} - \boldsymbol{X}\widehat{\boldsymbol{\beta}}_{\mathrm{R}}^{\mathrm{ML}}) = \boldsymbol{R}^\top \boldsymbol{W}(\widehat{\boldsymbol{\psi}}_{\mathrm{R}}^{\mathrm{ML}})(\boldsymbol{R}\widetilde{\boldsymbol{\beta}}(\widehat{\boldsymbol{\psi}}_{\mathrm{R}}^{\mathrm{ML}}) - \boldsymbol{r})$, which provides the score test. Thus, the Wald, LM, and LR test statistics are

$$\begin{aligned}
T_{\mathrm{W}} &= (\boldsymbol{R}\widetilde{\boldsymbol{\beta}}(\widehat{\boldsymbol{\psi}}^{\mathrm{ML}}) - \boldsymbol{r})^\top \boldsymbol{W}(\widehat{\boldsymbol{\psi}}^{\mathrm{ML}})(\boldsymbol{R}\widetilde{\boldsymbol{\beta}}(\widehat{\boldsymbol{\psi}}^{\mathrm{ML}}) - \boldsymbol{r}) = V(\widehat{\boldsymbol{\psi}}^{\mathrm{ML}}), \\
T_{\mathrm{LM}} &= (\boldsymbol{R}\widetilde{\boldsymbol{\beta}}(\widehat{\boldsymbol{\psi}}_{\mathrm{R}}^{\mathrm{ML}}) - \boldsymbol{r})^\top \boldsymbol{W}(\widehat{\boldsymbol{\psi}}_{\mathrm{R}}^{\mathrm{ML}})(\boldsymbol{R}\widetilde{\boldsymbol{\beta}}(\widehat{\boldsymbol{\psi}}_{\mathrm{R}}^{\mathrm{ML}}) - \boldsymbol{r}) = V(\widehat{\boldsymbol{\psi}}_{\mathrm{R}}^{\mathrm{ML}}), \\
T_{\mathrm{LR}} &= -2[\ell(\widehat{\boldsymbol{\beta}}_{\mathrm{R}}^{\mathrm{ML}}, \widehat{\boldsymbol{\psi}}_{\mathrm{R}}^{\mathrm{ML}}) - \ell(\widehat{\boldsymbol{\beta}}^{\mathrm{ML}}, \widehat{\boldsymbol{\psi}}^{\mathrm{ML}})],
\end{aligned} \tag{5.1}$$

where

$$V(\boldsymbol{\psi}) = (\boldsymbol{R}\widetilde{\boldsymbol{\beta}}(\boldsymbol{\psi}) - \boldsymbol{r})^\top \boldsymbol{W}(\boldsymbol{\psi})(\boldsymbol{R}\widetilde{\boldsymbol{\beta}}(\boldsymbol{\psi}) - \boldsymbol{r}).$$

The above test statistics are based on the ML estimators $\widehat{\boldsymbol{\beta}}^{\mathrm{ML}}$ and $\widehat{\boldsymbol{\psi}}^{\mathrm{ML}}$. However, there exist many other estimation methods for $\boldsymbol{\psi}$. For example, the ANOVA-type estimators, MINQU estimators, and restricted ML estimators are used in applications of mixed-effects models. Thus we need to extend the test statistics to the general consistent estimators of $\boldsymbol{\psi}$. Let $\widehat{\boldsymbol{\psi}}$ be an unrestricted consistent estimator of $\boldsymbol{\psi}$ which satisfies the following condition.

(T1) $\widehat{\boldsymbol{\psi}} = \widehat{\boldsymbol{\psi}}(\boldsymbol{y})$ be a general consistent estimator satisfying that $\widehat{\boldsymbol{\psi}}(-\boldsymbol{y}) = \widehat{\boldsymbol{\psi}}(\boldsymbol{y})$ and $\widehat{\boldsymbol{\psi}}(\boldsymbol{y} + \boldsymbol{X}\boldsymbol{\alpha}) = \widehat{\boldsymbol{\psi}}(\boldsymbol{y})$ for any p-dimensional vector $\boldsymbol{\alpha}$.

As a restricted estimator of $\boldsymbol{\psi}$, Kojima and Kubokawa (2013) suggested to use

$$\widehat{\boldsymbol{\psi}}_{\mathrm{R}} = \widehat{\boldsymbol{\psi}} - \boldsymbol{\Lambda}(\widehat{\boldsymbol{\psi}}) \begin{pmatrix} V_{(1)}(\widehat{\boldsymbol{\psi}}) \\ \vdots \\ V_{(q)}(\widehat{\boldsymbol{\psi}}) \end{pmatrix}, \tag{5.2}$$

where $\boldsymbol{\Lambda}(\boldsymbol{\psi}) = 2^{-1}\mathrm{E}[(\widehat{\boldsymbol{\psi}} - \boldsymbol{\psi})(\widehat{\boldsymbol{\psi}} - \boldsymbol{\psi})^\top]$ and $V_{(i)}(\boldsymbol{\psi}) = (\partial/\partial\psi_i)V(\boldsymbol{\psi})$. When $\widehat{\boldsymbol{\psi}}$ is the maximum likelihood (ML) estimator, $\widehat{\boldsymbol{\psi}}_{\mathrm{R}}$ is the restricted ML estimator under the constraint $\boldsymbol{R}\boldsymbol{\beta} = \boldsymbol{r}$. Define $\widehat{\boldsymbol{\beta}}$ and $\widehat{\boldsymbol{\beta}}_{\mathrm{R}}$ by $\widehat{\boldsymbol{\beta}} = \widetilde{\boldsymbol{\beta}}(\widehat{\boldsymbol{\psi}})$ and $\widehat{\boldsymbol{\beta}}_{\mathrm{R}} = \widetilde{\boldsymbol{\beta}}_{\mathrm{R}}(\widehat{\boldsymbol{\psi}}_{\mathrm{R}})$. Based on the general consistent estimators $\widehat{\boldsymbol{\psi}}$ and $\widehat{\boldsymbol{\psi}}_{\mathrm{R}}$, Kojima and Kubokawa (2013) proposed the Wald-type, LM-type, and modified LR-type test statistics, respectively, given by

$$\begin{aligned}
F_{\mathrm{W}} &= V(\widehat{\boldsymbol{\psi}}), \\
F_{\mathrm{LM}} &= V(\widehat{\boldsymbol{\psi}}_{\mathrm{R}}), \\
F_{\mathrm{mLR}} &= \{V(\widehat{\boldsymbol{\psi}}) + V(\widehat{\boldsymbol{\psi}}_{\mathrm{R}})\}/2.
\end{aligned} \tag{5.3}$$

The Bartlett corrections of the test statistics in (5.1) were derived by Rothenberg (1984) under the null and local alternative hypotheses. Using the same arguments, Kojima and Kubokawa (2013) derived the Bartlett-type corrections for the test statistics in (5.3). Let

$$
\begin{aligned}
a(\boldsymbol{\psi}) &= \sum_{i,j}^{q} \mathrm{tr}\,(\boldsymbol{\Sigma} \boldsymbol{A}_{(i)}^\top \boldsymbol{A}_{(j)}) \mathrm{Cov}_{H_0}(\widehat{\psi}_i, \widehat{\psi}_j), \\
b(\boldsymbol{\psi}) &= \frac{1}{2} \sum_{i,j}^{q} \left\{ \mathrm{tr}\,(\boldsymbol{B}_{(i)} \boldsymbol{B}_{(j)}) + \frac{1}{2} \mathrm{tr}\,(\boldsymbol{B}_{(i)}) \mathrm{tr}\,(\boldsymbol{B}_{(j)}) \right\} \mathrm{Cov}_{H_0}(\widehat{\psi}_i, \widehat{\psi}_j), \qquad (5.4) \\
c(\boldsymbol{\psi}) &= \sum_{i}^{q} \mathrm{tr}\,(\boldsymbol{B}_{(i)}) \mathrm{Bias}_{H_0}(\widehat{\psi}_i) + \sum_{i,j}^{q} \left\{ \frac{1}{2} \mathrm{tr}\,(\boldsymbol{B}_{(ij)}) + \mathrm{tr}\,(\boldsymbol{B}_{(i)} \boldsymbol{B}_{(j)}) \right\} \mathrm{Cov}_{H_0}(\widehat{\psi}_i, \widehat{\psi}_j),
\end{aligned}
$$

where $\mathrm{Cov}_{H_0}(\widehat{\psi}_i, \widehat{\psi}_j)$ and $\mathrm{Bias}_{H_0}(\widehat{\psi}_i)$ are the covariance of $\widehat{\psi}_i$ and $\widehat{\psi}_j$ and the bias of $\widehat{\psi}_i$ under the null hypothesis H_0, and

$$
\begin{aligned}
\boldsymbol{A}_{(i)} &= \boldsymbol{W}^{1/2} \boldsymbol{R} ((\boldsymbol{X}^\top \boldsymbol{\Sigma}^{-1} \boldsymbol{X})^{-1} \boldsymbol{X}^\top \boldsymbol{\Sigma}^{-1})_{(i)}, \\
\boldsymbol{B}_{(i)} &= -\boldsymbol{W}^{1/2} \boldsymbol{R} (\boldsymbol{X}^\top \boldsymbol{\Sigma}^{-1} \boldsymbol{X})^{-1}_{(i)} \boldsymbol{R}^\top \boldsymbol{W}^{1/2}, \\
\boldsymbol{B}_{(ij)} &= -\boldsymbol{W}^{1/2} \boldsymbol{R} (\boldsymbol{X}^\top \boldsymbol{\Sigma}^{-1} \boldsymbol{X})^{-1}_{(ij)} \boldsymbol{R}^\top \boldsymbol{W}^{1/2}.
\end{aligned}
$$

Let $\widehat{a}$, $\widehat{b}$, and $\widehat{c}$ be estimators given by $\widehat{a} = a(\widehat{\boldsymbol{\psi}})$, $\widehat{b} = b(\widehat{\boldsymbol{\psi}})$, and $\widehat{c} = c(\widehat{\boldsymbol{\psi}})$. Based on these functions, we obtain the test statistics with the Bartlett-type adjustments, given by

$$
\begin{aligned}
F_{\mathrm{W}}^{\mathrm{BC}} &= F_{\mathrm{W}} \Big(1 - \frac{1}{r} (\widehat{a} - \widehat{b} + \widehat{c}) - \frac{x}{r(r+2)} \widehat{b} \Big), \\
F_{\mathrm{LM}}^{\mathrm{BC}} &= F_{\mathrm{LM}} \Big(1 - \frac{1}{r} (-\widehat{a} - \widehat{b} + \widehat{c}) + \frac{x}{r(r+2)} \widehat{b} \Big), \qquad (5.5) \\
F_{\mathrm{mLR}}^{\mathrm{BC}} &= F_{\mathrm{LR}} \Big(1 - \frac{1}{r} (-\widehat{b} + \widehat{c}) \Big).
\end{aligned}
$$

Theorem 5.1 *Let $\widehat{\boldsymbol{\psi}}$ be a general consistent estimator satisfying condition* (T1). *Let $\widehat{\boldsymbol{\psi}}_{\mathrm{R}}$ be the restricted estimator defined in (5.2). Under appropriate conditions, the cumulative distribution functions of the Wald-type, ML-type, and modified LR-type test statistics with the Bartlett-type adjustments, given by* $\mathrm{P}(F_{\mathrm{W}}^{\mathrm{BC}} \le x)$, $\mathrm{P}(F_{\mathrm{LM}}^{\mathrm{BC}} \le x)$ *and* $\mathrm{P}(F_{\mathrm{mLR}}^{\mathrm{BC}} \le x)$, *are approximated as $F_r(x) + o(N^{-1})$ under the null hypothesis.*

For the proof and the conditions, see Kojima and Kubokawa (2013), who also provided the parametric bootstrap method for the Bartlett-type corrections.

We give an example of the Fay–Herriot model (4.1) where the normality is assumed. In this model, $\boldsymbol{\Sigma} = \psi \boldsymbol{I}_m + \boldsymbol{D}$. Let $\widehat{\psi}$ be a consistent estimator satisfying the condition in Theorem 5.1. From (5.2), the restricted estimator under H_0 is

$$
\widehat{\psi}_{\mathrm{R}} = \widehat{\psi} - \lambda(\widehat{\psi}) V_{(1)}(\widehat{\psi}),
$$

where $\lambda(\psi) = 2^{-1}\mathrm{E}[(\widehat{\psi} - \psi)^2]$ and

$$V_{(1)}(\psi) = -2(R\widetilde{\beta} - r)^\top WRH\Sigma^{-1}X(\widetilde{\beta}^* - \widetilde{\beta}) - (R\widetilde{\beta} - r)^\top WRHH^\top R^\top W(R\widetilde{\beta} - r),$$

for $\widetilde{\beta} = (X^\top\Sigma^{-1}X)^{-1}X^\top\Sigma^{-1}y$, $\widetilde{\beta}^* = (X^\top\Sigma^{-2}X)^{-1}X^\top\Sigma^{-2}y$,

$$W = [R(X^\top\Sigma^{-1}X)^{-1}R^\top]^{-1} \quad \text{and} \quad H = (X^\top\Sigma^{-1}X)^{-1}X^\top\Sigma^{-1}.$$

Also, $a(\psi)$, $b(\psi)$ and $c(\psi)$ in (5.4) are

$$\begin{aligned} a(\psi) =& \mathrm{tr}\,(WRHQH^\top R^\top)\mathrm{Var}_{H_0}(\widehat{\psi}),\\ b(\psi) =& \frac{1}{2}\Big[\mathrm{tr}\,\{(WRHH^\top R^\top)^2\} + \frac{1}{2}\{\mathrm{tr}\,(WRHH^\top R^\top)\}^2\Big]\mathrm{Var}_{H_0}(\widehat{\psi}),\\ c(\psi) =& \mathrm{tr}\,(WRHH^\top R^\top)\mathrm{Bias}_{H_0}(\widehat{\psi})\\ &+ \Big[\mathrm{tr}\,(WRHQH^\top R^\top) + \mathrm{tr}\,\{(WRHH^\top R^\top)^2\}\Big]\mathrm{Var}_{H_0}(\widehat{\psi}), \end{aligned}$$

for $Q = \Sigma^{-1} - \Sigma^{-1}X(X^\top\Sigma^{-1}X)^{-1}X^\top\Sigma^{-1}$. For the Prasad–Rao estimator in (4.14), we can obtain the Bartlett corrections by substituting $\mathrm{Bias}_{H_0}(\widehat{\psi}^{\mathrm{PR}}) = o(m^{-1})$ and $\mathrm{Var}_{H_0}(\widehat{\psi}^{\mathrm{PR}}) = (2/m^2)\mathrm{tr}\,(\Sigma^2) + o(m^{-1})$.

5.2 Information Criteria for Variable or Model Selection

Related to testing the hypothesis on the regression coefficients, the variable selection procedures are useful for choosing significant explanatory variables affecting the response variables.

The problem of selecting explanatory variables in linear mixed models is important and has been investigated in the literature. Of these, the *Akaike Information Criterion* (AIC), the *conditional Akaike Information Criterion* (cAIC), and the *Bayesian Information Criterion* (BIC) have been extensively studied and used in various applications. It is known that AIC and cAIC are criteria motivated from minimizing the prediction errors, but not consistent in the sense of selecting the true model. On the other hand, BIC is consistent, but it does not necessarily select the model which minimizes the prediction error. For a good account of the information criteria, see Konishi and Kitagawa (2007).

Consider the linear mixed model given in (2.1), where the normality is assumed for v and ϵ. For explaining the concepts of these criteria, let $f(y|v, \beta, \psi)$ and $f(v|\psi)$ be the conditional density of y given v and the marginal density of v, respectively, where $y|v \sim \mathrm{N}(X\beta + Zv, R_e(\psi))$ and $v \sim \mathrm{N}(0, R_v(\psi))$. Then, the marginal density of y is written by $f_m(y|\beta, \psi) = \int f(y|v, \beta, \psi)f(v|\psi)\mathrm{d}v$, which has the marginal distribution $\mathrm{N}(X\beta, \Sigma(\psi))$.

The AIC proposed by Akaike (1973, 1974) is based on the thought of choosing a model which minimizes an unbiased estimator of the expected Kullback–Leibler information. Let $\widehat{\boldsymbol{\beta}}(\boldsymbol{y})$ and $\widehat{\boldsymbol{\psi}}(\boldsymbol{y})$ be consistent estimators of $\boldsymbol{\beta}$ and $\boldsymbol{\psi}$ based on $\boldsymbol{y}$. The expected Kullback–Leibler information is defined by

$$R(\boldsymbol{\beta}, \boldsymbol{\psi}; \widehat{\boldsymbol{\beta}}, \widehat{\boldsymbol{\psi}}) = \mathrm{E}_y\Big[\int \Big(\log \frac{f_m(\boldsymbol{y}^*|\boldsymbol{\beta}, \boldsymbol{\psi})}{f_m(\boldsymbol{y}^*|\widehat{\boldsymbol{\beta}}(\boldsymbol{y}), \widehat{\boldsymbol{\psi}}(\boldsymbol{y}))}\Big) f_m(\boldsymbol{y}^*|\boldsymbol{\beta}, \boldsymbol{\psi})\mathrm{d}\boldsymbol{y}^*\Big],$$

which can be interpreted as a risk function for estimating $(\boldsymbol{\beta}, \boldsymbol{\psi})$ by $(\widehat{\boldsymbol{\beta}}, \widehat{\boldsymbol{\psi}})$ relative to the Kullback–Leibler distance. This quantity measures the prediction error in predicting future variable $\boldsymbol{y}^*$ based on the model $f_m(\boldsymbol{y}^*|\widehat{\boldsymbol{\beta}}(\boldsymbol{y}), \widehat{\boldsymbol{\psi}}(\boldsymbol{y}))$. In this sense, AIC is a criterion of finding a model which can provide a good prediction in light of minimizing the prediction error. $R(\boldsymbol{\beta}, \boldsymbol{\psi}; \widehat{\boldsymbol{\beta}}, \widehat{\boldsymbol{\psi}})$ is rewritten as

$$\begin{aligned}&\iint \{\log f_m(\boldsymbol{y}^*|\boldsymbol{\beta}, \boldsymbol{\psi})\} f_m(\boldsymbol{y}^*|\boldsymbol{\beta}, \boldsymbol{\psi})\mathrm{d}\boldsymbol{y}^* f_m(\boldsymbol{y}|\boldsymbol{\beta}, \boldsymbol{\psi})\mathrm{d}\boldsymbol{y} \\ &\quad - \iint \{\log f_m(\boldsymbol{y}^*|\widehat{\boldsymbol{\beta}}(\boldsymbol{y}), \widehat{\boldsymbol{\psi}}(\boldsymbol{y}))\} f_m(\boldsymbol{y}^*|\boldsymbol{\beta}, \boldsymbol{\psi})\mathrm{d}\boldsymbol{y}^* f_m(\boldsymbol{y}|\boldsymbol{\beta}, \boldsymbol{\psi})\mathrm{d}\boldsymbol{y}.\end{aligned}$$

Since the first term is irrelevant to the model $f_m(\boldsymbol{y}^*|\widehat{\boldsymbol{\beta}}(\boldsymbol{y}), \widehat{\boldsymbol{\psi}}(\boldsymbol{y}))$, it is sufficient to estimate the second term. Thus, the *Akaike Information* (AI) is defined by

$$AI = -2\iint \{\log f_m(\boldsymbol{y}^*|\widehat{\boldsymbol{\beta}}(\boldsymbol{y}), \widehat{\boldsymbol{\psi}}(\boldsymbol{y}))\} f_m(\boldsymbol{y}^*|\boldsymbol{\beta}, \boldsymbol{\psi}) f_m(\boldsymbol{y}|\boldsymbol{\beta}, \boldsymbol{\psi})\mathrm{d}\boldsymbol{y}^*\mathrm{d}\boldsymbol{y},$$

and AIC is derived as an asymptotically unbiased estimator of AI, namely, $\mathrm{E}[AIC] = AI + o(1)$. When AIC is given as an exact unbiased estimator of AI, it is called the exact AIC, which was suggested by Sugiura (1978), but, in general, it is difficult to get the exact AIC in LMM.

It is noted that AIC stated above is based on the marginal distribution of $\boldsymbol{y}$, namely, it measures the prediction error of the predictor based on the marginal distribution $\mathrm{N}(\boldsymbol{X}\boldsymbol{\beta}, \boldsymbol{\Sigma})$. This means that the marginal AIC is not appropriate for the focus on the prediction of specific areas or random effects as explained in the context of the small area estimation. Taking this point into account, Vaida and Blanchard (2005) proposed the conditional AIC as an asymptotically unbiased estimator of cAI, where cAI is the conditional Akaike information defined by

$$\begin{aligned}cAI = -2\iiint &\log\{f(\boldsymbol{y}^*|\widetilde{\boldsymbol{v}}(\boldsymbol{y}), \widehat{\boldsymbol{\beta}}(\boldsymbol{y}), \widehat{\boldsymbol{\psi}}(\boldsymbol{y}))\} \\ &f(\boldsymbol{y}^*|\boldsymbol{v}, \boldsymbol{\beta}, \boldsymbol{\psi}) f(\boldsymbol{y}|\boldsymbol{v}, \boldsymbol{\beta}, \boldsymbol{\psi}) f(\boldsymbol{v}|\boldsymbol{\psi})\mathrm{d}\boldsymbol{y}^*\mathrm{d}\boldsymbol{y}\mathrm{d}\boldsymbol{v},\end{aligned}$$

where $\widehat{\boldsymbol{v}}(\boldsymbol{y}) = \boldsymbol{R}_v(\widehat{\boldsymbol{\psi}})\boldsymbol{Z}^\top\boldsymbol{\Sigma}(\widehat{\boldsymbol{\psi}})^{-1}\{\boldsymbol{y} - \boldsymbol{X}\widehat{\boldsymbol{\beta}}(\boldsymbol{y})\}$ is the empirical Bayes estimator of $\boldsymbol{v}$. When $\boldsymbol{\psi}$ is known, Vaida and Blanchard (2005) derived an exact unbiased estimator of cAI, and it gives the same value as DIC, the deviance information criterion proposed by Spiegelhalter et al. (2002) for Bayesian inference.

Since it is difficult to obtain exact unbiased estimators of AI and cAI in general linear mixed models, we provide their asymptotic approximations. Hereafter, the regression coefficient $\boldsymbol{\beta}$ is estimated by the GLS $\widehat{\boldsymbol{\beta}} = \widetilde{\boldsymbol{\beta}}(\widehat{\boldsymbol{\psi}})$ for $\widetilde{\boldsymbol{\beta}}(\boldsymbol{\psi})$ in (2.12). For estimator $\widehat{\boldsymbol{\psi}}$ of $\boldsymbol{\psi}$, we assume (T1) and (T2) $\widehat{\boldsymbol{\psi}}$ is approximated as $\widehat{\boldsymbol{\psi}} = \boldsymbol{\psi} + \widetilde{\boldsymbol{\psi}} + O_p(N^{-1})$ for $\widetilde{\boldsymbol{\psi}} = (\widetilde{\psi}_1, \ldots, \widetilde{\psi}_q)^\top = O_p(N^{-1/2})$.

These conditions are satisfied by the ML-, REML-, and ANOVA-type estimators.

Theorem 5.2 *Under the conditions* (T1), (T2), *and additional appropriate conditions, the expectation of* $-2\log f_m(\boldsymbol{y}|\widehat{\boldsymbol{\beta}}, \widehat{\boldsymbol{\psi}})$ *is approximated as* $\mathrm{E}[-2\log f_m(\boldsymbol{y}|\widehat{\boldsymbol{\beta}}, \widehat{\boldsymbol{\psi}})] = AI(\boldsymbol{\psi}) - 2p - 2h_m(\boldsymbol{\psi}) + o(1)$, *where*

$$h_m(\boldsymbol{\psi}) = \frac{1}{2}\sum_{i=1}^{q} E\big[\mathrm{tr}\,\big(\boldsymbol{\Sigma}_{(i)}\boldsymbol{\nabla}_y\boldsymbol{\nabla}_y^\top\widetilde{\psi}_i\big)\big]. \tag{5.6}$$

For the proof and the conditions, see Kubokawa (2011). Since the AIC is an asymptotically unbiased estimator of $AI(\boldsymbol{\psi})$, from (5.6) in Theorem 5.2, the AIC is given by

$$\mathrm{AIC} = -2\log f_m(\boldsymbol{y}|\widehat{\boldsymbol{\beta}}, \widehat{\boldsymbol{\psi}}) + 2p + 2h_m(\widehat{\boldsymbol{\psi}}), \tag{5.7}$$

where $-2\log f_m(\boldsymbol{y}|\widehat{\boldsymbol{\beta}}, \widehat{\boldsymbol{\psi}}) = N\log(2\pi) + \log|\widehat{\boldsymbol{\Sigma}}| + (\boldsymbol{y} - \boldsymbol{X}\widehat{\boldsymbol{\beta}})^\top\widehat{\boldsymbol{\Sigma}}^{-1}(\boldsymbol{y} - \boldsymbol{X}\widehat{\boldsymbol{\beta}})$ for $\widehat{\boldsymbol{\Sigma}} = \boldsymbol{\Sigma}(\widehat{\boldsymbol{\psi}})$.

It is noted that $2p$ is the penalty term corresponding to the estimation of $\boldsymbol{\beta}$ and $2h_m(\boldsymbol{\psi})$ is the penalty arisen from estimating unknown parameters $\boldsymbol{\psi}$. When we treat the estimators $\widehat{\boldsymbol{\psi}}$ given in (2.20) or (2.22), from (2.36), $\widehat{\boldsymbol{\psi}}$ can be approximated as $\widehat{\boldsymbol{\psi}} = \boldsymbol{\psi} + \boldsymbol{A}^{-1}\mathbf{col}_a(\ell_a) + O_p(N^{-1})$. Since $\widetilde{\boldsymbol{\psi}} = \boldsymbol{A}^{-1}\mathbf{col}_a(\ell_a)$ or $\widetilde{\psi}_i = \sum_{j=1}^q(\boldsymbol{A})^{ij}\ell_j$, we have $\boldsymbol{\nabla}_y\boldsymbol{\nabla}_y^\top\widetilde{\psi}_i = 2\sum_{j=1}^q(\boldsymbol{A})^{ij}\boldsymbol{C}_j$, which implies that $\sum_{i=1}^q \mathrm{tr}\,\big(\boldsymbol{\Sigma}_{(i)}\boldsymbol{\nabla}_y\boldsymbol{\nabla}_y^\top\widetilde{\psi}_i\big) = 2\sum_{i=1}^q\sum_{j=1}^q(\boldsymbol{A})^{ij}\mathrm{tr}\,(\boldsymbol{\Sigma}_{(i)}\boldsymbol{C}_j) = 2q$. Thus, one gets $h_m(\boldsymbol{\psi}) = q$, the dimension of $\boldsymbol{\psi}$, which corresponds to the AIC originated from Akaike (1973). This means that AIC is useful for selecting the parameters $\boldsymbol{\psi}$ as well as selecting the variables in $\boldsymbol{X}$.

Concerning the cAIC, define $\Delta_c(\boldsymbol{\psi})$ by

$$\begin{aligned}\Delta_c(\boldsymbol{\psi}) =& -2\rho(\boldsymbol{\psi}) + \sum_{i=1}^{q}\mathrm{tr}\,\big(\boldsymbol{\Sigma}(\boldsymbol{\Sigma}^{-1}\boldsymbol{R}_e\boldsymbol{\Sigma}^{-1})_{(i)}\boldsymbol{\Sigma}\mathrm{E}[\boldsymbol{\nabla}_y\boldsymbol{\nabla}_y^\top\widetilde{\psi}_i]\big)\\ &+2\sum_{i=1}^{q}\mathrm{tr}\,\big(\boldsymbol{R}_e\{(\boldsymbol{\Sigma}^{-1})_{(i)} - (\boldsymbol{R}_e^{-1})_{(i)}\}\big)\mathrm{Bias}(\widehat{\psi}_i)\\ &+\sum_{i=1}^{q}\sum_{j=1}^{q}\mathrm{tr}\,\big(\boldsymbol{R}_e\{(\boldsymbol{\Sigma}^{-1})_{(ij)} - (\boldsymbol{R}_e^{-1})_{(ij)}\}\big)\mathrm{Cov}(\widehat{\psi}_i, \widehat{\psi}_j),\end{aligned}$$

where $\rho(\boldsymbol{\psi})$ is the effective degrees of freedom defined by

$$\rho(\boldsymbol{\psi}) = \mathrm{tr}\,[(\boldsymbol{X}^\top\boldsymbol{\Sigma}^{-1}\boldsymbol{X})^{-1}\boldsymbol{X}^\top\boldsymbol{\Sigma}^{-1}\boldsymbol{R}_e\boldsymbol{\Sigma}^{-1}\boldsymbol{X}] + N - \mathrm{tr}\,(\boldsymbol{R}_e\boldsymbol{\Sigma}^{-1}).$$

Theorem 5.3 *Under the conditions* (T1), (T2), *and additional appropriate conditions, it holds that* $\mathrm{E}[-2\log f(\boldsymbol{y}|\widehat{\boldsymbol{v}},\widehat{\boldsymbol{\beta}},\widehat{\boldsymbol{\psi}})] = cAI(\boldsymbol{\psi}) + \Delta_c(\boldsymbol{\psi}) + o(1)$*, where* $\widehat{\boldsymbol{v}} = \widehat{\boldsymbol{R}}_v\boldsymbol{Z}^\top\widehat{\boldsymbol{\Sigma}}^{-1}(\boldsymbol{y} - \boldsymbol{X}\widehat{\boldsymbol{\beta}})$ *for* $\widehat{\boldsymbol{R}}_v = \boldsymbol{R}_v(\widehat{\boldsymbol{\psi}})$.

Since the term $\mathrm{tr}\,(\boldsymbol{R}_e\boldsymbol{\Sigma}^{-1})$ in the function $\rho(\boldsymbol{\psi})$ is of order $O(N)$, it is noted that $\Delta_c(\widehat{\boldsymbol{\psi}})$ is not an asymptotically unbiased estimator of $\Delta_c(\boldsymbol{\psi})$ up to $O(1)$, namely, $\mathrm{E}[\Delta_c(\widehat{\boldsymbol{\psi}})] \neq \Delta_c(\boldsymbol{\psi}) + o(1)$. Thus, we need to approximate the expectation $\mathrm{E}[\mathrm{tr}\,(\boldsymbol{R}_e(\widehat{\boldsymbol{\psi}})\boldsymbol{\Sigma}(\widehat{\boldsymbol{\psi}})^{-1})]$ up to $O(1)$. Based on this approximation, we define the function $h_c(\boldsymbol{\psi})$ by

$$
\begin{aligned}
h_c(\boldsymbol{\psi}) = &-\frac{1}{2}\sum_{i=1}^{q}\mathrm{tr}\left(\boldsymbol{\Sigma}(\boldsymbol{\Sigma}^{-1}\boldsymbol{R}_e\boldsymbol{\Sigma}^{-1})_{(i)}\boldsymbol{\Sigma}\mathrm{E}[\nabla_{\boldsymbol{y}}\nabla_{\boldsymbol{y}}^\top\widetilde{\psi}_i]\right)\\
&-\sum_{i=1}^{q}\mathrm{tr}\,[(\boldsymbol{R}_e)_{(i)}(\boldsymbol{R}_e^{-1} - \boldsymbol{\Sigma}^{-1})]\mathrm{Bias}(\widehat{\psi}_i)\\
&-\frac{1}{2}\sum_{i=1}^{q}\sum_{j=1}^{q}\mathrm{tr}\,[(\boldsymbol{R}_e)_{(ij)}(\boldsymbol{R}_e^{-1} - \boldsymbol{\Sigma}^{-1})]\mathrm{Cov}(\widehat{\psi}_i,\widehat{\psi}_j)\\
&-\sum_{i=1}^{q}\sum_{j=1}^{q}\mathrm{tr}\,[(\boldsymbol{R}_e)_{(i)}\{(\boldsymbol{R}_e^{-1})_{(j)} - (\boldsymbol{\Sigma}^{-1})_{(j)}\}]\mathrm{Cov}(\widehat{\psi}_i,\widehat{\psi}_j).
\end{aligned}
\tag{5.8}
$$

Although the expression in (5.8) seems complicated, it can be simplified for specific cases. In the case that $\boldsymbol{R}_e$ is a linear function of $\boldsymbol{\psi}$ and $\widehat{\boldsymbol{\psi}}$ is second-order unbiased, $h_c(\boldsymbol{\psi})$ is written as

$$
\begin{aligned}
h_c(\boldsymbol{\psi}) = &-\frac{1}{2}\sum_{i=1}^{q}\mathrm{tr}\left(\boldsymbol{\Sigma}(\boldsymbol{\Sigma}^{-1}\boldsymbol{R}_e\boldsymbol{\Sigma}^{-1})_{(i)}\boldsymbol{\Sigma}\mathrm{E}[\nabla_{\boldsymbol{y}}\nabla_{\boldsymbol{y}}^\top\widetilde{\psi}_i]\right)\\
&-\sum_{i=1}^{q}\sum_{j=1}^{q}\mathrm{tr}\,[(\boldsymbol{R}_e)_{(i)}\{(\boldsymbol{R}_e^{-1})_{(j)} - (\boldsymbol{\Sigma}^{-1})_{(j)}\}]\mathrm{Cov}(\widehat{\psi}_i,\widehat{\psi}_j).
\end{aligned}
$$

Theorem 5.4 *Under the conditions* (T1), (T2), *and additional appropriate conditions, the estimator* $-2\rho(\widehat{\boldsymbol{\psi}}) - 2h_c(\widehat{\boldsymbol{\psi}})$ *is an asymptotically unbiased estimator of* $\Delta_c(\boldsymbol{\psi})$*, namely,* $\mathrm{E}[-2\rho(\widehat{\boldsymbol{\psi}}) - 2h_c(\widehat{\boldsymbol{\psi}})] = \Delta_c(\boldsymbol{\psi}) + o(1)$.

Since the cAIC is an asymptotically unbiased estimator of $cAI(\boldsymbol{\psi})$, from Theorems 5.3 and 5.4, the cAIC is given by

$$
\mathrm{cAIC} = -2\log f(\boldsymbol{y}|\widehat{\boldsymbol{v}},\widehat{\boldsymbol{\beta}},\widehat{\boldsymbol{\psi}}) + 2\rho(\widehat{\boldsymbol{\psi}}) + 2h_c(\widehat{\boldsymbol{\psi}}),
\tag{5.9}
$$

where $-2\log f(\boldsymbol{y}|\widehat{\boldsymbol{v}},\widehat{\boldsymbol{\beta}},\widehat{\boldsymbol{\psi}}) = N\log(2\pi) + \log|\widehat{\boldsymbol{R}}_e| + (\boldsymbol{y} - \boldsymbol{X}\widehat{\boldsymbol{\beta}} - \boldsymbol{Z}\widehat{\boldsymbol{v}})^\top\widehat{\boldsymbol{R}}_e^{-1}(\boldsymbol{y} - \boldsymbol{X}\widehat{\boldsymbol{\beta}} - \boldsymbol{Z}\widehat{\boldsymbol{v}})$.

It is noted that $h_c(\boldsymbol{\psi})$ is equal to $h_m(\boldsymbol{\psi})$ if $\boldsymbol{\Sigma} = \boldsymbol{R}_e$, namely $\boldsymbol{Z} = \boldsymbol{0}$. Vaida and Blanchard (2005) treated the case of known $\boldsymbol{\psi}$ and showed that the penalty term is asymptotically given by $2\rho(\boldsymbol{\psi})$. These observations suggest that $2\rho(\boldsymbol{\psi})$ is a penalty term for selecting variables in $\boldsymbol{X}$ for fixed $\boldsymbol{\psi}$ and $2h_c(\boldsymbol{\psi})$ may be connected to penalty arisen from estimating unknown $\boldsymbol{\psi}$.

When estimators $\widehat{\boldsymbol{\psi}}$ are given in (2.20) or (2.22), from (2.36), $\widetilde{\boldsymbol{\psi}}$ is $\widetilde{\boldsymbol{\psi}} = \boldsymbol{A}^{-1}\mathbf{col}_a(\ell_a)$ or $\widetilde{\psi}_i = \sum_{j=1}^q (\boldsymbol{A})^{ij}\ell_j$. Then, we have $\nabla_y \nabla_y^\top \widetilde{\psi}_i = 2\sum_{j=1}^q (\boldsymbol{A})^{ij}\boldsymbol{C}_j$, which implies that

$$-\frac{1}{2}\sum_{i=1}^q \mathrm{tr}\,[\boldsymbol{\Sigma}(\boldsymbol{\Sigma}^{-1}\boldsymbol{R}\boldsymbol{\Sigma}^{-1})_{(i)}\boldsymbol{\Sigma}\nabla_y\nabla_y^\top \widetilde{\psi}_i]$$

$$= \sum_{i=1}^q \sum_{j=1}^q (\boldsymbol{A})^{ij}[2\mathrm{tr}\,(\boldsymbol{\Sigma}_{(i)}\boldsymbol{R}_e\boldsymbol{C}_j) - \mathrm{tr}\,\{(\boldsymbol{R}_e)_{(i)}\boldsymbol{C}_j\}].$$

For the Fay–Herriot model with $\boldsymbol{\Sigma} = \psi \boldsymbol{I}_m + \boldsymbol{D}$, we have

$$h_c(\psi) = 2\frac{\mathrm{tr}\,(\boldsymbol{D}\boldsymbol{P}\boldsymbol{W}_1\boldsymbol{P})}{\mathrm{tr}\,(\boldsymbol{P}\boldsymbol{W}_1)} = 2\frac{\mathrm{tr}\,(\boldsymbol{D}\boldsymbol{W}_1)}{\mathrm{tr}\,(\boldsymbol{W}_1)} + o(1),$$

for $\boldsymbol{P} = \boldsymbol{I} - \boldsymbol{X}(\boldsymbol{X}^\top\boldsymbol{X})^{-1}\boldsymbol{X}^\top$. This value is $2\mathrm{tr}\,(\boldsymbol{D})/m$ for the Prasad–Rao estimator, $2\mathrm{tr}\,(\boldsymbol{D}\boldsymbol{\Sigma}^{-1})/\mathrm{tr}\,(\boldsymbol{\Sigma}^{-1})$ for the Fay–Herriot estimator, and $2\mathrm{tr}\,(\boldsymbol{D}\boldsymbol{\Sigma}^{-2})/\mathrm{tr}\,(\boldsymbol{\Sigma}^{-2})$ for the REML estimator.

The Bayesian information criterion (BIC) proposed by Schwarz (1978) assumes a proper prior distribution $\pi(\boldsymbol{\beta}, \boldsymbol{\psi})$ formally and evaluate asymptotically the marginal distribution

$$f_\pi(\boldsymbol{y}) = \iint f_m(\boldsymbol{y}|\boldsymbol{\beta}, \boldsymbol{\psi})\pi(\boldsymbol{\beta}, \boldsymbol{\psi})\mathrm{d}\boldsymbol{\beta}\mathrm{d}\boldsymbol{\psi}.$$

The Laplace approximation is used to approximate it as $-2\log\{f_\pi(\boldsymbol{y})\} = \mathrm{BIC} + o_p(\log(N))$, where

$$\mathrm{BIC} = -2\log\{f_m(\boldsymbol{y}|\widehat{\boldsymbol{\beta}}, \widehat{\boldsymbol{\psi}})\} + (p+q)\log(N), \tag{5.10}$$

where $-2\log\{f_m(\boldsymbol{y}|\widehat{\boldsymbol{\beta}}, \widehat{\boldsymbol{\psi}})\}$ is given below (5.7). The distinction between AIC and BIC appears in the penalty terms as seen from (5.7) and (5.10). The Bayesian criteria like Bayes factors use all the prior information on $(\boldsymbol{\beta}, \boldsymbol{\psi})$, while all the prior information is neglected in BIC, because the prior information comes into neglected terms asymptotically.

There are some other procedures suggested in linear mixed models. Jiang et al. (2008) proposed the Fance methods for selecting explanatory variables and random effects. Kawakubo et al. (2018) derived the conditional Akaike information criterion under covariate shift when the values of covariates in the model for prediction differ from those in the model for observed data. Müller et al. (2013) provided a good survey on model selection in linear mixed models.

References

Akaike H (1973) Information theory and an extension of the maximum likelihood principle. In: Petrov BN, Csaki F (eds) 2nd international symposium on information theory. Budapest, Akademia Kiado, pp 267–281

Akaike H (1974) A new look at the statistical model identification. System identification and time-series analysis. IEEE Trans Autom Contr AC-19:716–723

Jiang J, Rao JS, Gu Z, Nguyen T (2008) Fence methods for mixed model selection. Ann Statist 36:1669–1692

Kojima M, Kubokawa T (2013) Bartlett-type adjustments for hypothesis testing in linear models with general error covariance matrices. J Multivariate Anal 122:162–174

Kawakubo Y, Sugasawa S, Kubokawa T (2018) Conditional Akaike information under covariate shift with application to small area estimation. Canadian J Statist 46:316–335

Konishi S, Kitagawa G (2007) Information criteria and statistical modeling. Springer, Berlin

Kubokawa T (2011) Conditional and unconditional methods for selecting variables in linear mixed models. J Multivariate Anal 102:641–660

Müller S, Scealy JL, Welsh AH (2013) Model selection in linear mixed models. Statist Sci 28:135–167

Rothenberg T (1984) Hypothesis testing in linear models when the error covariance matrix is nonscalar. Econometrika 52:827–842

Schwarz G (1978) Estimating the dimension of a model. Ann Statist 6:461–464

Spiegelhalter DJ, Best NG, Carlin BP, van der Linde A (2002) Bayesian measures of model complexity and fit (with Discussion). J Royal Statist Soc B 64:583–639

Sugiura N (1978) Further analysis of the data by Akaike's information criterion and the finite corrections. Commun Statist—Theory Methods 1:13–26

Vaida F, Blanchard S (2005) Conditional Akaike information for mixed-effects models. Biometrika 92:351–370

Chapter 6
Advanced Theory of Basic Small Area Models

In Chap. 4, we introduced two famous small area models, Fay–Herriot and nested error regression models, and provided basic theory of parameter estimation and EBLUP. However, the standard methods have some problems in practical use. We here illustrate advanced methodology and theory for the basic small area models.

6.1 Adjusted Likelihood Methods

6.1.1 Strictly Positive Estimate of Random Effect Variance

Consider the Fay–Herriot model (4.1) with normality assumptions, $\varepsilon_i \sim \mathrm{N}(0, D_i)$ and $v_i \sim \mathrm{N}(0, A)$. Remember that the BLUP of θ_i is

$$\widetilde{\theta}_i = \boldsymbol{x}_i^\top \widehat{\boldsymbol{\beta}} + \frac{A}{A + D_i}(y_i - \boldsymbol{x}_i^\top \widehat{\boldsymbol{\beta}}),$$

where the random effects variance A determines the amount of shrinkage. Typically, the variance A is estimated by maximizing the profile likelihood given by

$$L_{\mathrm{P}}(A) = c|\boldsymbol{\Sigma}|^{-1/2} \exp\left(-\frac{1}{2}\boldsymbol{y}^\top \boldsymbol{P}\boldsymbol{y}\right)$$

and the residual likelihood given by

$$L_{\mathrm{R}}(A) = c|\boldsymbol{X}^\top \boldsymbol{\Sigma}^{-1}\boldsymbol{X}|^{-1/2}|\boldsymbol{\Sigma}|^{1/2} \exp\left(-\frac{1}{2}\boldsymbol{y}^\top \boldsymbol{P}\boldsymbol{y}\right),$$

S. Sugasawa and T. Kubokawa, *Mixed-Effects Models and Small Area Estimation*, JSS Research Series in Statistics, https://doi.org/10.1007/978-981-19-9486-9_6

where c is a generic constant that does not depend on A and $\boldsymbol{P} = \boldsymbol{\Sigma}^{-1} - \boldsymbol{\Sigma}^{-1}\boldsymbol{X}(\boldsymbol{X}^\top\boldsymbol{\Sigma}^{-1}\boldsymbol{X})^{-1}\boldsymbol{X}^\top\boldsymbol{\Sigma}^{-1}$ and $\boldsymbol{\Sigma} = \mathrm{diag}(D_1, \ldots, D_m) + AI_m$. A practical problem in estimating A is that the estimator of A obtained by maximizing $L_{\mathrm{P}}(A)$ or $L_{\mathrm{RE}}(A)$ might be 0, especially when m is small, under which the BLUP reduces to $\widetilde{\theta}_i = \boldsymbol{x}_i^\top\widehat{\boldsymbol{\beta}}$. This is not plausible since heterogeneity among areas cannot be taken into account. To overcome the difficulty, Li and Lahiri (2010) introduced a modification of the likelihood function called an adjusted likelihood, given by

$$L_{\mathrm{ad}}(A) = h(A)L(A), \tag{6.1}$$

where $L(A)$ is a given likelihood function and $h(A)$ is an adjustment term. Then, the adjusted likelihood estimator of A is defined as $\widehat{A}_{\mathrm{ad}} = \mathrm{argmax}_A L_{\mathrm{ad}}(A)$. It is desirable that $h(A)$ down-weights the likelihood function around the small value of A. Specifically, Li and Lahiri (2010) proposed the adjustment function, $h(A) = A$. In this case, it follows that $L_{\mathrm{ad}}(0) = 0$ for $L(A) = L_{\mathrm{P}}(A)$ or $L_{\mathrm{R}}(A)$. Moreover, it can be shown that $\lim_{A\to\infty} AL_{\mathrm{P}}(A) \to 0$ for $m > 2$ and $\lim_{A\to\infty} AL_{\mathrm{R}}(A) \to 0$ for $m > p + 2$. Since $AL_{\mathrm{P}}(A)$ and $AL_{\mathrm{R}}(A)$ are strictly positive when $m \geq 2$ and $m \geq p + 2$, respectively, and are continuous functions of A. Hence, under the conditions, the estimator $\widehat{A}_{ad}$ is positive.

Under some suitable conditions, it holds that

$$\mathrm{E}[(\widehat{A}_{\mathrm{ad}} - A)^2] = \frac{2}{\mathrm{tr}(\boldsymbol{\Sigma}^{-2})} + o(m^{-1}),$$

for both $\widehat{A}_{\mathrm{ad}}$ with profile and residual likelihood, and

$$\mathrm{E}[\widehat{A}_{\mathrm{ad}}] - A = \frac{\mathrm{tr}(\boldsymbol{P} - \boldsymbol{\Sigma}^{-1}) + 2/A}{\mathrm{tr}(\boldsymbol{\Sigma}^{-2})} + o(m^{-1})$$

for the adjusted profile likelihood estimator and

$$\mathrm{E}[\widehat{A}_{\mathrm{ad}}] - A = \frac{2/A}{\mathrm{tr}(\boldsymbol{\Sigma}^{-2})} + o(m^{-1})$$

for the adjusted residual likelihood estimator. Thus, the adjusted likelihood estimators are consistent for large m. Moreover, the asymptotic variance for $\widehat{A}_{\mathrm{ad}}$ is the same as one for non-adjusted profile (or residual) likelihood estimator, which means that positiveness can be assured without loose efficiency under large m. of A and $h(A)$ is an adjustment factor. On the other hand, the asymptotic biases of the adjusted profile (or residual) likelihood estimators are of order $O(m^{-1})$, which are the same as the order of the non-adjusted profile likelihood estimator, but higher than that of the non-adjusted residual likelihood estimator.

Based on the properties, it would be natural to seek another form of the adjustment term $h(A)$ to have better asymptotic properties. In what follows, we focus only on the adjusted restricted likelihood estimator. Yoshimori and Lahiri (2014a) derived unified asymptotic results under general form of $h(A)$, given by

$$\mathrm{E}[\widehat{A}_{\mathrm{ad}}] - A = \frac{2\widetilde{l}_{\mathrm{ad}}^{(1)}}{\mathrm{tr}(\boldsymbol{\Sigma}^{-2})} + o(m^{-1}),$$

where $\widetilde{l}_{\mathrm{ad}}^{(1)} = \mathrm{d}\log h(A)/\mathrm{d}A$, and the asymptotic variance is the same regardless of the form of $h(A)$. The asymptotic bias formula given in Li and Lahiri (2010) can be obtained by $\mathrm{d}\log A/\mathrm{d}A = 1/A$. To make the bias of the adjusted residual likelihood estimator of $o(m^{-1})$ like the non-adjusted residual likelihood estimator, Yoshimori and Lahiri (2014a) proposed the following form of the adjustment term:

$$h_{\mathrm{YL}}(A) = [\tan^{-1}\{\mathrm{tr}(\boldsymbol{I}_m - \boldsymbol{B})\}]^{1/m},$$

where $\boldsymbol{B} \equiv \boldsymbol{B}(A) = \mathrm{diag}(B_1, \ldots, B_m)$ and $B_i = D_i/(A + D_i)$. The straightforward calculation shows that

$$\widetilde{l}_{\mathrm{ad}}^{(1)} = \frac{1}{m} \frac{\sum_{i=1}^m \{D_i/(A + D_i)^2\}}{\tan^{-1}\{\mathrm{tr}(\boldsymbol{I}_m - \boldsymbol{B})\}\{1 + \mathrm{tr}(I - B)^2\}} = O(m^{-1}),$$

so that the asymptotic bias of the adjusted residual likelihood estimator with $h_{\mathrm{YL}}(A)$ is of $o(m^{-1})$, the same order of the non-adjusted residual likelihood estimator. Since $h_{\mathrm{YL}}(0) = 0$ and $L_{\mathrm{R}}(A) < \infty$, it follows that $h_{\mathrm{YL}}(0)L_{\mathrm{R}}(0) = 0$. Moreover, using the property $0 < h_{\mathrm{YL}}(A) < \pi/2$ for $A > 0$, it holds that $\lim_{A\to\infty} h_{\mathrm{YL}} L_{\mathrm{R}}(A) = 0$ for $m > p$. Hence, the estimator is strictly positive.

6.1.2 Adjusted Likelihood for Empirical Bayes Confidence Intervals

Such an adjusted likelihood method has been adopted not only for avoiding zero estimate but also for constructing confidence intervals. Consider empirical Bayes confidence intervals of θ_i under the Fay–Herriot model. Under the Fay–Herriot model, the conditional distribution of θ_i given y_i is $\mathrm{N}(\widetilde{\theta}_i(\boldsymbol{\beta}, A), \sigma_i^2(A))$, where $\widetilde{\theta}_i(\boldsymbol{\beta}, A) = (1 - B_i)y_i + B_i\boldsymbol{x}_i^\top\boldsymbol{\beta}$ and $\sigma_i^2(A) = AD_i/(A + D_i)$. Then, empirical Bayes confidence interval of θ_i is given by

$$I_i^{\mathrm{EB}}(\widehat{\boldsymbol{\beta}}, \widehat{A}) : \widetilde{\theta}_i(\widehat{\boldsymbol{\beta}}, \widehat{A}) \pm z_{\alpha/2}\sigma_i(\widehat{A}),$$

where $z_{\alpha/2}$ the upper $100(1 - \alpha/2)$ point of $\mathrm{N}(0, 1)$, and $\widehat{\boldsymbol{\beta}}$ and $\widehat{A}$ are some estimators of $\boldsymbol{\beta}$ and A. It is known that the coverage accuracy is generally of $O(m^{-1})$, that is, $\mathrm{P}(\theta_i \in I_i^{\mathrm{EB}}(\widehat{\boldsymbol{\beta}}, \widehat{A})) = 1 - \alpha + O(m^{-1})$, as shown in, for example, Chatterjee et al. (2008). This is not accurate enough in practice when m is not so large.

To solve the property, let $\widehat{A}_{h_i}$ be the adjusted residual likelihood estimator with an area-wise adjustment function $h_i(A)$, namely, $Ah_{h_i} = \text{argmax} h_i(A)L_{\text{R}}(A)$. Under some regularity conditions, Yoshimori and Lahiri (2014b) derived the following asymptotic expansion of the coverage probability of $I_i^{\text{EB}}(\widehat{\boldsymbol{\beta}}, \widehat{A}_{h_i})$. We first note that

$$\text{P}(\theta_i \in I_i^{\text{EB}}(\widehat{\boldsymbol{\beta}}, \widehat{A}_{h_i})) = 1 - \alpha + z\phi(z)\frac{a_i + b_i h_i(A)}{m} + O(m^{-3/2}), \tag{6.2}$$

where

$$a_i = -\frac{m}{\text{tr}\left(V^{-2}\right)}\left[\frac{4D_i}{A\,(A+D_i)^2} + \frac{\left(1+z^2\right)D_i^2}{2\,A^2\,(A+D_i)^2}\right] - \frac{mD_i}{A\,(A+D_i)}\boldsymbol{x}_i^\top \text{Var}(\widetilde{\boldsymbol{\beta}}(A))\boldsymbol{x}_i,$$

$$b_i \equiv b_i\left[h_i(A)\right] = \frac{2m}{\text{tr}\left(\boldsymbol{\Sigma}^{-2}\right)}\frac{D_i}{A\,(A+D_i)} \times \widetilde{l}_{i;\text{ad}}^{(1)}$$

and $\widetilde{l}_{i;\text{ad}}^{(1)} = \text{d}\log h_i(A)/\text{d}A$, noting that $h_i(A)$ has appeared in the coverage error only through its derivative $\widetilde{l}_{i;\text{ad}}^{(1)}$. Here $\widetilde{\boldsymbol{\beta}}(A)$ is an estimator of $\boldsymbol{\beta}$, and we consider two estimators, generalized least square and ordinary least square estimators. From expression (6.2), it seems possible to reduce the coverage error to the order $O(m^{-3/2})$ by choosing $h_i(A)$ such that the order $O(m^{-1})$ term in the right-hand side of (6.2) vanishes. Specifically, we obtain an expression for $h_i(A)$ by solving the following differential equation:

$$a_i + b_i[h_i(A)] = 0, \quad i = 1, \ldots, m,$$

and then obtain the adjusted residual likelihood estimator $\widehat{A}_{h_i}$, which is used to construct the accurate empirical Bayes confidence intervals satisfying

$$\text{P}(\theta_i \in I_i^{\text{EB}}(\widehat{\boldsymbol{\beta}}, \widehat{A}_{h_i})) = 1 - \alpha + O(m^{-3/2}).$$

As shown in Yoshimori and Lahiri (2014a), the solution of $h_i(A)$ under the generalized least squares estimator of $\boldsymbol{\beta}$ does not have a closed-form expression, but the solution under ordinary least squares estimator of $\boldsymbol{\beta}$ is obtained as

$$\begin{aligned} h_i(A) =& C A^{(1/4)(1+z^2)}\,(A+D_i)^{(1/4)(7-z^2)}\left[\prod_{i=1}^{m}(A+D_i)\right]^{(1/2)q_i} \\ & \times \exp\left\{-\frac{1}{2}\text{tr}\left(\boldsymbol{\Sigma}^{-1}\right)\boldsymbol{x}_i^\top\left(\boldsymbol{X}^\top\boldsymbol{X}\right)^{-1}\boldsymbol{X}^\top\boldsymbol{\Sigma}\boldsymbol{X}\left(\boldsymbol{X}^\top\boldsymbol{X}\right)^{-1}\boldsymbol{x}_i\right\}, \end{aligned}$$

where $q_i = \boldsymbol{x}_i^\top\left(\boldsymbol{X}^\top\boldsymbol{X}\right)^{-1}\boldsymbol{x}_i$ and C is a generic constant that does not depend on A. Furthermore, under the balanced case, $D_i = D$ for $i = 1, \ldots, m$, we have a more simplified expression

$$h_i(A) = CA^{(1/4)(1+z^2)}(A + D_i)^{(1/4)(7-z^2)+mq_i/2}$$

and the resulting estimator $\widehat{A}_{h_i}$ is unique if $m > (4 + p)/(1 - q_i)$.

The required condition for the existence of $\widehat{A}_{h_i}$ could be restrictive when at least one leverage value q_i is high. To overcome the problem, Hirose (2017) developed an alternative adjustment approach. We start with the following general form of the confidence interval:

$$I_i^*(\widehat{\boldsymbol{\beta}}, \widehat{A}_{h_i}) : \widetilde{\theta}_i(\widehat{\boldsymbol{\beta}}, \widehat{A}_{h_i}) \pm z_{\alpha/2} s_i(\widehat{A}_{h_i}, c_i^*),$$

where $s_i^2(A, c_i^*) = g_{1i}(A) + g_{2i}(A) + c_i^* g_{3i}(A)$ and

$$g_{1i}(A) = \frac{AD_i}{A + D_i}, \quad g_{2i}(A) = B_i^2 \boldsymbol{x}_i^\top \left(\boldsymbol{X}^\top \boldsymbol{\Sigma}^{-1} \boldsymbol{X}\right)^{-1} \boldsymbol{x}_i, \quad g_{3i}(A) = \frac{2B_i^2}{(A + D_i)\mathrm{tr}(\boldsymbol{\Sigma}^{-2})}.$$

Note that $s_i^2(A, 2)$ corresponds to the second-order approximation of the MSE of $\widetilde{\theta}_i(\widehat{\boldsymbol{\beta}}, \widehat{A})$. For the confidence interval, the condition for the adjustment term to ensure that the interval $I_i^*(\widehat{\boldsymbol{\beta}}, \widehat{A}_{h_i})$ is second-order correct is given by

$$\widetilde{l}_{i,\mathrm{ad}}^{(1)}(A) = \frac{7 - z_{\alpha/2}^2 - 4c_i^*}{4(A + D_i)} + \frac{1 + z_{\alpha/2}^2}{4A} + O\left(m^{-1/2}\right).$$

Based on the results, Hirose (2017) suggested setting c_i^* and $h_i(A)$ as

$$c_i^* = c_{\mathrm{NAS}}^* \equiv \frac{7 - z_{\alpha/2}^2}{4}, \qquad h_i(A) = h_{\mathrm{NAS}}(A) \equiv A^{(1+z_{\alpha/2}^2)/4},$$

so that the adjustment term h_i does not depend on i. Hence, the resulting confidence interval is

$$I_i^*(\widehat{\boldsymbol{\beta}}, \widehat{A}_{\mathrm{NAS}}) : \widetilde{\theta}_i(\widehat{\boldsymbol{\beta}}, \widehat{A}_{\mathrm{NAS}}) - z_{\alpha/2} s_i(\widehat{A}_{\mathrm{NAS}}, c_{\mathrm{NAS}}^*),$$

where $\widehat{A}_{\mathrm{NAS}} = \mathrm{argmax}_A h_{\mathrm{NAS}}(A) L_R(A)$ is the non-area-specific adjusted residual likelihood estimator. It is also proved that $\widehat{A}_{\mathrm{NAS}}$ exist when $m > p + (1 + z_{\alpha/2}^2)/2$, which is a much milder condition than one required in the area-specific adjustment given in Yoshimori and Lahiri (2014b).

Example 6.1 (*Batting average data*) Hirose (2017) compared some empirical Bayes confidence intervals by using batting average data provided by Efron and Morris (1975). Let y_i be a batting average and x_i be the previous seasonal batting average for ith player. By applying sin–arcsin transformation for both y_i and x_i, a balanced case $D_i = 1$ is considered. It should be noted that the maximum leverage value is 0.79, which leads to failure of condition to ensure the existence of area-wise adjusted likelihood estimator by Yoshimori and Lahiri (2014a). Instead, non-area-specific adjusted likelihood method by Hirose (2017) can be applied. In addition, the direct interval, $y_i \pm z_{\alpha/2}\sqrt{D_i}$, and the standard empirical Bayes confidence interval are applied. The results indicated that the length of $I_i^*(\widehat{\boldsymbol{\beta}}, \widehat{A}_{\rm NAS})$ is always less than that of the direct interval for all the players. Moreover, all true seasonal batting averages were included in the interval $I_i^*(\widehat{\boldsymbol{\beta}}, \widehat{A}_{\rm NAS})$, while they were not in the other intervals. See Hirose (2017) for more detailed results.

6.1.3 *Adjusted Likelihood for Solving Multiple Small Area Estimation Problems*

The adjusted likelihood is also useful for solving other problems in small area estimation. Hirose and Lahiri (2018) proposed a new but simple area-wise adjustment factor $h_{i0}(A) = A + D_i$ to achieve several desirable properties simultaneously. One of the important properties is the second-order unbiasedness of the MSE estimator. Specifically, let

$$\widehat{A}_{i,\rm MG} = \text{argmax}_A h_{i0}(A) h_{\rm YL}(A) L_{\rm R}(A),$$

where $h_{\rm YL}(A)$ is the adjustment term given in Sect. 6.1.1, and define the MSE estimator as

$$\widehat{M}_{i,\rm MG} = g_{1i}(\widehat{A}_{i,\rm MG}) + g_{2i}(\widehat{A}_{i,\rm MG}) + g_{3i}(\widehat{A}_{i,\rm MG}).$$

Then, it holds that $\mathrm{E}[\widehat{M}_{i,\rm MG}] = \mathrm{E}[\{\widetilde{\theta}_i(\widehat{\boldsymbol{\beta}}, \widehat{A}_{i,\rm MG}) - \theta_i\}^2] + o(m^{-1})$. Hirose (2019) generalized the approach by considering a more general form of the MSE estimator given by

$$\widehat{M}_{i,\rm G} = g_{1i}(\widehat{A}_{i,\rm G}) + g_{2i}(\widehat{A}_{i,\rm G}) + c_i g_{3i}(\widehat{A}_{i,\rm G}),$$

with c_i satisfying $c_i \geq -q_i/2$, where $\widehat{A}_{i,\rm G}$ is a general adjusted likelihood estimator

$$\widehat{A}_{i,\rm G} = \text{argmax}_A (A + D_i)^{2-c_i} h_{\rm YL}(A) L_{\rm R}(A).$$

Hirose (2019) proved that $\widehat{M}_{i,\rm G}$ is the second-order unbiased estimator of the MSE for general c_i, which includes the results in Hirose and Lahiri (2018) as a special case where $c_i = 1$.

6.2 Observed Best Prediction

The classical FH model (4.1) implicitly assumes that the regression part $\boldsymbol{x}_i^\top \boldsymbol{\beta}$ is correctly specified, and the estimation of model parameters including $\boldsymbol{\beta}$ as well as the prediction of θ_i is carried out under the assumed model. However, any assumed model is subject to model misspecification. Jiang et al. (2011) considered the situation where the true model is $\theta_i = \mu_i + v_i$ with $v_i \sim \mathrm{N}(0, A)$ and μ_i being the true mean. Note that μ_i is not necessarily equivalent to $\boldsymbol{x}_i^\top \boldsymbol{\beta}$. Then, they focused on a reasonable estimation method for regression coefficients $\boldsymbol{\beta}$ under possible model misspecification. To this end, they considered the total mean squared prediction error (MSPE) of the best predictor of θ_i, namely $\widetilde{\theta}_i(\boldsymbol{\beta}, A)$ minimizing $\mathrm{E}[(\widetilde{\theta}_i - \theta_i)^2]$ over all predictor θ_i if the assumed Fay–Herriot model (4.1) is correct and the true parameter is given. Let $\boldsymbol{\Gamma} = \mathrm{diag}(1 - B_1, \ldots, 1 - B_m)$, then the best predictor $\widetilde{\boldsymbol{\theta}}(\boldsymbol{\beta}, A) \equiv (\widetilde{\theta}_1(\boldsymbol{\beta}, A), \ldots, \widetilde{\theta}_m(\boldsymbol{\beta}, A))$ can be expressed as

$$\widetilde{\boldsymbol{\theta}}(\boldsymbol{\beta}, A) = \boldsymbol{y} - \boldsymbol{\Gamma}(\boldsymbol{y} - \boldsymbol{X}\boldsymbol{\beta}).$$

The MSPE of $\widetilde{\boldsymbol{\theta}}(\boldsymbol{\beta}, A)$ is given by

$$\mathrm{MSPE}(\widetilde{\boldsymbol{\theta}}(\boldsymbol{\beta}, A)) = \mathrm{E}[\{\boldsymbol{y} - \boldsymbol{\Gamma}(\boldsymbol{y} - \boldsymbol{X}\boldsymbol{\beta}) - \boldsymbol{\theta}\}^\top\{\boldsymbol{y} - \boldsymbol{\Gamma}(\boldsymbol{y} - \boldsymbol{X}\boldsymbol{\beta}) - \boldsymbol{\theta}\}] = I_1 + I_2,$$

where

$$\begin{aligned} I_1 &= \mathrm{E}[(\boldsymbol{y} - \boldsymbol{X}\boldsymbol{\beta})^\top\boldsymbol{\Gamma}^2(\boldsymbol{y} - \boldsymbol{X}\boldsymbol{\beta})], \\ I_2 &= \mathrm{E}[(\boldsymbol{y} - \boldsymbol{\theta})^\top(\boldsymbol{y} - \boldsymbol{\theta})] - 2\mathrm{E}[(\boldsymbol{y} - \boldsymbol{\theta})^\top\boldsymbol{\Gamma}(\boldsymbol{y} - \boldsymbol{X}\boldsymbol{\beta})]. \end{aligned}$$

Since

$$I_2 = \mathrm{tr}(\boldsymbol{D}) - 2\mathrm{tr}(\boldsymbol{\Gamma}\boldsymbol{D}) = 2A\mathrm{tr}(\boldsymbol{\Gamma}) - \mathrm{tr}(\boldsymbol{D}),$$

for $\boldsymbol{D} = \mathrm{diag}(D_1, \ldots, D_m)$, an unbiased estimator of the MSPE can be obtained. Hence, the objective function for $\boldsymbol{\beta}$ and A is

$$Q(\boldsymbol{\beta}, A) = (\boldsymbol{y} - \boldsymbol{X}\boldsymbol{\beta})^\top\boldsymbol{\Gamma}^2(\boldsymbol{y} - \boldsymbol{X}\boldsymbol{\beta}) + 2A\mathrm{tr}(\boldsymbol{\Gamma}).$$

This expression suggests that the minimizer of $\boldsymbol{\beta}$ is obtained as

$$\widetilde{\boldsymbol{\beta}} = (\boldsymbol{X}^\top\boldsymbol{\Gamma}^2\boldsymbol{X})^{-1}\boldsymbol{X}^\top\boldsymbol{\Gamma}^2\boldsymbol{y}, \tag{6.3}$$

which is called the observed best predictive (OBP) estimator of $\boldsymbol{\beta}$. Note that expression (6.3) is different from the maximum likelihood estimator or GLS estimator, $(\boldsymbol{X}^\top\boldsymbol{\Sigma}^{-1}\boldsymbol{X})^{-1}\boldsymbol{X}^\top\boldsymbol{\Sigma}^{-1}\boldsymbol{y}$. Jiang et al. (2011) also developed general OBP estimators under linear mixed models and asymptotic theory of OBP estimators.

Sugasawa et al. (2019) considered a selection criteria based on the OBP estimator in the Fay–Herriot model. Let (j) be a candidate model, so that the assumed mean model is $\boldsymbol{X}_{(j)}\boldsymbol{\beta}_{(j)}$. The OBP estimator of $\boldsymbol{\beta}_{(j)}$ and associated predictor of $\boldsymbol{\theta}$ are

$$\widehat{\boldsymbol{\beta}}_{(j)\mathrm{OBP}} = (\boldsymbol{X}_{(j)}^\top\boldsymbol{\Gamma}^2\boldsymbol{X}_{(j)})^{-1}\boldsymbol{X}_{(j)}^\top\boldsymbol{\Gamma}^2\boldsymbol{y}, \quad \widehat{\boldsymbol{\theta}}_{(j)} = \boldsymbol{y} - \boldsymbol{\Gamma}(\boldsymbol{y} - \boldsymbol{X}_{(j)}\widehat{\boldsymbol{\beta}}_{(j)\mathrm{OBP}}).$$

The MSPE of $\widehat{\boldsymbol{\theta}}_{(j)}$ is expressed as

$$\mathrm{E}[(\widehat{\boldsymbol{\theta}}_{(j)} - \boldsymbol{\theta})^\top(\widehat{\boldsymbol{\theta}}_{(j)} - \boldsymbol{\theta})] \equiv I_1 - 2I_2 + I_3,$$

where

$$I_1 = \mathrm{E}\left[(\boldsymbol{y}-\boldsymbol{\theta})^\top(\boldsymbol{y}-\boldsymbol{\theta})\right], \quad I_2 = \mathrm{E}\left[(\boldsymbol{y}-\boldsymbol{\theta})^\top\boldsymbol{\Gamma}\left(\boldsymbol{y}-\boldsymbol{X}_{(j)}\widehat{\boldsymbol{\beta}}_{(j)\mathrm{OBP}}\right)\right],$$
$$I_3 = \mathrm{E}\left[\left(\boldsymbol{y}-\boldsymbol{X}_{(j)}\widehat{\boldsymbol{\beta}}_{(j)\mathrm{OBP}}\right)^\top\boldsymbol{\Gamma}^2\left(\boldsymbol{y}-\boldsymbol{X}_{(j)}\widehat{\boldsymbol{\beta}}_{(j)\mathrm{OBP}}\right)\right].$$

It holds that $I_1 = \mathrm{tr}(\boldsymbol{D})$ and $I_2 = \mathrm{tr}\{(\boldsymbol{I}_m - \boldsymbol{P}_{(j)}\boldsymbol{\Gamma}\boldsymbol{D})\}$, where $\boldsymbol{P}_{(j)} = \boldsymbol{\Gamma}\boldsymbol{X}_{(j)}(\boldsymbol{X}_{(j)^\top}\boldsymbol{\Gamma}^2\boldsymbol{X}_{(j)})^{-1}\boldsymbol{X}_{(j)}{}^\top\boldsymbol{\Gamma}$. Hence, minimizing the unbiased estimator of the MSPE with respect to the model index (j) is equivalent to minimizing the following criteria:

$$C(j) = \left(\boldsymbol{y}-\boldsymbol{X}_{(j)}\widehat{\boldsymbol{\beta}}_{(j)\mathrm{OBP}}\right)^\top\boldsymbol{\Gamma}^2\left(\boldsymbol{y}-\boldsymbol{X}_{(j)}\widehat{\boldsymbol{\beta}}_{(j)\mathrm{OBP}}\right) + 2\mathrm{tr}(\boldsymbol{P}_{(j)}\boldsymbol{\Gamma}\boldsymbol{D}). \tag{6.4}$$

The second term is regarded as a penalty function for the complexity of the candidate model (j). In fact, under unbalanced case, $D_i = D$, it reduces to $\mathrm{tr}(\boldsymbol{P}_{(j)}\boldsymbol{\Gamma}\boldsymbol{D}) = D^2 p_j/(A+D)$, where p_j is the rank of $\boldsymbol{X}_{(j)}$. Note that the random effect variance A in the criteria is estimated from the full model. We define the best model as

$$\widehat{j} = \mathrm{argmin}_j C(j),$$

and the resulting estimator of $\boldsymbol{\beta}$ is

$$\widehat{\beta}_{\{j\}k} = \begin{cases} \widehat{\beta}_{(j)k} & \text{if } k \in I_j, \\ 0 & \text{if } k \notin I_j, \end{cases},$$

where I_j is an index set corresponding to the model (j). The estimator $\widehat{\beta}_{\{j\}k}$ is called observed best selective predictive (OBSP) estimator. Under some regularity conditions, it can be shown that $\sup_{1\le k\le p}\mathrm{E}\left\{\left(\widehat{\beta}_{(j\}k} - \beta_k^*\right)^2\right\} = O\left(m^{-1}\right)$ as $m \to \infty$.

Example 6.2 (*Variable selection for estimating household expenditure in Japan*) Sugasawa et al. (2019) applied the selection criteria (6.4) in the Fay–Herriot model for estimating the household expenditure on education per month in $m = 47$ prefectures in Japan. The response variable y_i $(i = 1, \ldots, m)$ is the log of areal mean of household expenditure, and the following four covariates are used: areal mean of household expenditure on education ('edc'), population ('pop'), proportion of the people under 15 years old ('young'), and proportion of the labor force ('labor'). Among all the combinations of the four covariates, the new selection criteria (6.4) selected a model with 'edc' and 'young', while both AIC and BIC based on the marginal likelihood selected a model with only 'edc'. Such inconsistency of the results would come from the different philosophy of constructing the selection criteria.

6.3 Robust Methods

In practice, outliers are often contained in data. The model assumption such as normality can be violated for such outlying data, and the inclusion of outliers would significantly affect estimation of model parameters. We here review robust methods against outliers under unit-level and area-level models.

6.3.1 *Unit-Level Models*

For estimating population parameters in the framework of finite population, existing outliers in the sampled data can invalidate the parametric assumptions of the nested error regression model (4.19). To take account of existence of outliers, Sinha and Rao (2009) robustify the likelihood equations of the linear mixed models. Let $\boldsymbol{y}_i = (y_{i1}, \ldots, y_{in_i})$ and $\boldsymbol{X}_i = (\boldsymbol{x}_{i1}, \ldots, \boldsymbol{x}_{in_i})$ be a vector of response variables and a covariate matrix in the ith area. Remember that the nested error regression model is given by

$$\boldsymbol{y}_i = \boldsymbol{X}_i\boldsymbol{\beta} + \mathbf{1}_{n_i} v_i + \boldsymbol{\epsilon}_i, \quad i = 1, \ldots, m,$$

where $v_i \sim \mathrm{N}(0, \tau^2)$ and $\boldsymbol{\epsilon}_i \sim \mathrm{N}(0, \sigma^2 \boldsymbol{I}_{n_i})$. Note that $\mathrm{Var}(\boldsymbol{y}_i) \equiv \boldsymbol{\Sigma}_i = \tau^2 \boldsymbol{J}_{n_i} + \sigma^2 \boldsymbol{I}_{n_i}$ for $\boldsymbol{J}_{n_i} = \mathbf{1}_{n_i}\mathbf{1}_{n_i}^\top$. The robust estimating equation for v_i is

$$\sigma^{-1}\mathbf{1}_{n_i}^\top \boldsymbol{\psi}_K(\sigma^{-1}(\boldsymbol{y}_i - \boldsymbol{X}_i\boldsymbol{\beta} - \mathbf{1}_{n_i} v_i)) - \tau^{-1}\psi_K(\tau^{-1}v_i) = 0, \tag{6.5}$$

where $\psi_K(t) = u\min(1, K/|u|)$ is Huber's ψ-function with a tuning constant $K > 0$, and $\boldsymbol{\psi}_K(t) = (\psi_K(t_1), \ldots, \psi_K(t_{n_i}))$. When there are some outlying observations such that the absolute standardized residual $\sigma^{-1}|y_{ij} - \boldsymbol{x}_{ij}^\top\boldsymbol{\beta} - v_i|$ is large, the standard equation is highly affected by such outliers. However, in the robust Eq. (6.5), the residual is cut off at K owing to the ψ-function, so that the outliers are less effective in the equation. Therefore, K is a key tuning parameter that controls the degree of robustness of Eq. (6.5) and a common choice is $K = 1.345$. Furthermore, the robust Eq. (6.5) reduces to the standard (non-robust) equation for v_i when $K \to \infty$. Unlike the standard equation for v_i, the solution of (6.5) cannot be obtained in a closed form, but the robust equation can be numerically solved by a Newton–Raphson method.

The same idea can be applied in the estimation of the unknown parameters. The robust likelihood equations for $\boldsymbol{\beta}$, σ^2, and τ^2 are given by

$$\sum_{i=1}^{m} X_i^\top \Sigma_i^{-1} \psi_K(r_i) = 0,$$

$$\sum_{i=1}^{m} \left\{ (\tau^2 + \sigma^2) \psi_K(r_i)^\top \Sigma_i^{-2} \psi_K(r_i) - \mathrm{tr}(c\Sigma_i^{-1}) \right\} = 0,$$

$$\sum_{i=1}^{m} \left\{ (\tau^2 + \sigma^2) \psi_K(r_i)^\top \Sigma_i^{-1} J_{n_i} \Sigma_i^{-1} \psi_K(r_i) - \mathrm{tr}(c_K \Sigma_i^{-1} J_{n_i}) \right\} = 0,$$

where $r_i = (y_i - X_i\beta)/\sqrt{\sigma^2 + \tau^2}$ and $c_K = \mathrm{E}[\{\psi_K(U)\}^2]$ for $U \sim \mathrm{N}(0, 1)$. The above equations can be solved by a Newton–Raphson method. Let $\widehat{\beta}_R, \widehat{\sigma}_R^2$, and $\widehat{\tau}_R^2$ be the robust estimators as the solution of the above equation, and $\widetilde{v}_i^R(\beta, \sigma^2, \tau^2)$ be robust random effect estimate as the solution of (6.5). Then, given the covariates for non-sampled units, the robust predictor (called REBLUP) of the population mean is given by

$$\frac{1}{N_i}\left[\sum_{j=1}^{n_i} y_{ij} + \sum_{j=n_i+1}^{N_i} \{x_{ij}^\top \widehat{\beta}_R + \widetilde{v}_i^R(\widehat{\beta}_R, \widehat{\sigma}_R^2, \widehat{\tau}_R^2)\}\right].$$

However, the above estimator assumes that the non-sampled values in the population are drawn from a distribution with the same mean as the sampled non-outliers, which may be unrealistic. An improved version of the robust estimator of the population mean is given in, for example, Dongmo-Jiongo et al. (2013); Chambers et al. (2014).

Example 6.3 (*County crop areas*) Battese et al. (1988) considered the estimation of areas under corn and soybeans for $m = 12$ counties (regarded as 'small areas') in North-Central Iowa, based on farm-interview survey data and satellite pixel data. Areas of corn and soybeans were obtained in 37 sample segments from the 12 counties. The dataset contains the number of segments in each county, the number of hectares of corn and soybeans for each sample segment, the number of pixels classified by the LANDSAT satellite as corn and soybeans for each sample segment, and the mean number of pixels per segment in each county classified as corn and soybeans. Battese et al. (1988) identified one observation in Hardin county to be an influential outlier, and they simply deleted this observation when predicting the corn and soybeans areas based on the nested error regression model. Sinha and Rao (2009) applied the robust method to the data to explore the ability of handling such influential observations. The maximum likelihood estimates of the variance parameters are $\widehat{\tau}^2 = 47.8$ and $\widehat{\sigma}^2 = 280.2$ while the robust estimates are $\widehat{\tau}_R^2 = 102.7$ and $\widehat{\sigma}_R^2 = 225.6$, so that σ^2 seems to be over-estimated due to the outlier. Furthermore, both REBLUP and EBLUP provide similar predictions for most of the counties, but in the Hardin County (where an outlier exists), EBLUP is 131.3 and REBLUP is 136.9.

6.3.2 Area-Level Models

In the context of the Fay–Herriot model (4.1), the EBLUP or empirical Bayes estimator may over- or under-shrink the direct estimator y_i toward the regression estimator $\boldsymbol{x}_i^\top\widehat{\boldsymbol{\beta}}$ when the outlying areas are included.

To see the effect of each observation, Ghosh et al. (2008) investigated the influence function of each observation $(y_i, \boldsymbol{x}_i)$ on the posterior distribution of $\boldsymbol{\beta}$. We first introduce the following divergence measure between two density functions, f_1 and f_2:

$$D_\lambda(f_1, f_2) = \frac{1}{\lambda(\lambda+1)}\int\left\{\left(\frac{f_1(x)}{f_2(x)}\right)^\lambda - 1\right\} f_1(x)dx.$$

In particular, when f_1 and f_2 are density functions of $\mathrm{N}(\boldsymbol{\mu}_1, \boldsymbol{\Sigma}_1)$ and $\mathrm{N}(\boldsymbol{\mu}_2, \boldsymbol{\Sigma}_2)$, respectively, it follows that

$$\begin{aligned} D_\lambda(f_1, f_2) = &\frac{1}{\lambda(\lambda+1)}\left(\exp\left[\frac{\lambda(\lambda+1)}{2}(\boldsymbol{\mu}_1 - \boldsymbol{\mu}_2)^\top\{(1+\lambda)\boldsymbol{\Sigma}_2 - \lambda\boldsymbol{\Sigma}_1\}^{-1}(\boldsymbol{\mu}_1 - \boldsymbol{\mu}_2)\right]\right. \\ &\left.\times |\boldsymbol{\Sigma}_1|^{-\lambda/2}|\boldsymbol{\Sigma}_2|^{-(\lambda-1)/2}|(1+\lambda)\boldsymbol{\Sigma}_2 - \lambda\boldsymbol{\Sigma}_1|^{1/2} - 1\right). \end{aligned}$$

Note that $D_\lambda(f_1, f_2)$ is one to one with $(\boldsymbol{\mu}_1 - \boldsymbol{\mu}_2)^\top\{(1+\lambda)\boldsymbol{\Sigma}_2 - \lambda\boldsymbol{\Sigma}_1\}^{-1}(\boldsymbol{\mu}_1 - \boldsymbol{\mu}_2)$. We apply the above result to compare the posterior distributions of $\boldsymbol{\beta}$ based on all the observations and observations without ith area. Given the uniform prior $\pi(\boldsymbol{\beta}) \propto 1$, the posterior distribution of $\boldsymbol{\beta}$ is $\boldsymbol{\beta}|\boldsymbol{y}, \boldsymbol{X} \sim \mathrm{N}(\boldsymbol{\mu}_1, \boldsymbol{\Sigma}_1)$, where

$$\boldsymbol{\mu}_1 = (\boldsymbol{X}^\top\boldsymbol{\Sigma}^{-1}\boldsymbol{X})^{-1}\boldsymbol{X}^\top\boldsymbol{\Sigma}^{-1}\boldsymbol{y}, \qquad \boldsymbol{\Sigma}_1 = (\boldsymbol{X}\boldsymbol{\Sigma}^{-1}\boldsymbol{X})^{-1},$$

where $\boldsymbol{y} = (y_1, \dots, y_m)$, $\boldsymbol{X} = (\boldsymbol{x}_1, \dots, \boldsymbol{x}_m)$ and $\boldsymbol{\Sigma} = \mathrm{diag}(A + D_1, \dots, A + D_m)$. On the other hand, the posterior distribution of $\boldsymbol{\beta}$ given $y_{(-i)}$ (removing $\boldsymbol{y}$ from y_i) and $\boldsymbol{X}_{(-i)}$ (removing $\boldsymbol{X}$ from $\boldsymbol{x}_i$) is $\boldsymbol{\beta}|\boldsymbol{y}_{(-i)}, \boldsymbol{X}_{(-i)} \sim \mathrm{N}(\boldsymbol{\mu}_2, \boldsymbol{\Sigma}_2)$, where

$$\boldsymbol{\mu}_2 = (\boldsymbol{X}_{(-i)}^\top\boldsymbol{\Sigma}_{(-i)}^{-1}\boldsymbol{X}_{(-i)})^{-1}\boldsymbol{X}_{(-i)}^\top\boldsymbol{\Sigma}_{(-i)}^{-1}\boldsymbol{y}_{(-i)}, \qquad \boldsymbol{\Sigma}_2 = (\boldsymbol{X}_{(-i)}\boldsymbol{\Sigma}_{(-i)}^{-1}\boldsymbol{X}_{(-i)})^{-1}),$$

and $\boldsymbol{\Sigma}_{(-i)}$ is a diagonal matrix similar to $\boldsymbol{\Sigma}$ with the ith diagonal element removed. By the straightforward calculation, it can be shown that $D_\lambda(f_1, f_2)$ is one to one with $(\boldsymbol{\mu}_1 - \boldsymbol{\mu}_2)^\top\{(1+\lambda)\boldsymbol{\Sigma}_2 - \lambda\boldsymbol{\Sigma}_1\}^{-1}(\boldsymbol{\mu}_1 - \boldsymbol{\mu}_2)$ is a quadratic form in $y_i - \boldsymbol{x}_i^\top\widetilde{\boldsymbol{\beta}}$ with $\widetilde{\boldsymbol{\beta}} = (\boldsymbol{X}^\top\boldsymbol{\Sigma}^{-1}\boldsymbol{X})^{-1}\boldsymbol{X}^\top\boldsymbol{\Sigma}^{-1}\boldsymbol{y}$. Hence, it would be reasonable to restrict the amount of shrinkage by controlling the residuals $y_i - \boldsymbol{x}_i^\top\widetilde{\boldsymbol{\beta}}$. To make the residuals scale free, we standardize them by s_i, where

$$s_i^2(A) \equiv \mathrm{Var}(y_i - \boldsymbol{x}_i^\top\widetilde{\boldsymbol{\beta}}) = A + D_i - \boldsymbol{x}_i^\top(\boldsymbol{X}^\top\boldsymbol{\Sigma}^{-1}\boldsymbol{X})^{-1}\boldsymbol{x}_i.$$

Based on these considerations, Ghosh et al. (2008) proposed the robust BLUP or Bayes estimator of θ_i as

$$\widetilde{\theta}_i^{\rm RB} = y_i - \frac{D_i s_i(A)}{A + D_i} \psi_K \left(\frac{y_i - \boldsymbol{x}_i^\top \widetilde{\boldsymbol{\beta}}}{s_i(A)} \right), \quad i = 1, \ldots, m,$$

where $\psi_K(t)$ is Huber's ψ-function.

The MSE of $\widetilde{\theta}_i^{\rm RB}$ is

$$\mathrm{E}[(\widetilde{\theta}_i^{\rm RB} - \theta_i)^2] = \mathrm{E}[(\widetilde{\theta}_i - \theta_i)^2] + \frac{2D_i^4}{(A + D_i)^2} \left\{ (1 + K^2)\Phi(-K) - K\Phi(K) \right\},$$

where $\widetilde{\theta}_i$ is the (non-robust) Bayes estimator obtained by $K \to \infty$ in $\widetilde{\theta}_i^{\rm RB}$. Hence, the second term corresponds to MSE inflation (i.e., loss of efficiency) due to the use of finite K. Thus, K is determined based on a trade-off between the large deviations $(y_i - \boldsymbol{x}_i^\top \widetilde{\boldsymbol{\beta}})/s_i(A)$ and the excess MSE that one is willing to tolerate when the assumed model is true. One possible way is to set a tolerate percentage α of MSE (e.g., $\alpha = 0.05$ or 0.1), and determine K by solving the equation $\mathrm{E}[(\widetilde{\theta}_i^{\rm RB} - \theta_i)^2]/\mathrm{E}[(\widetilde{\theta}_i - \theta_i)^2] = 1 + \alpha$.

As an alternative approach, Sinha and Rao (2009) modified the equation to obtain BLUP of θ_i using Huber's ψ-function. The proposed robust BLUP is the solution of the equation

$$D_i^{-1/2} \psi_K \left\{ D_i^{-1/2}(y_i - \theta_i) \right\} - A^{-1/2} \psi_K \left\{ A^{-1/2}(\theta_i - \boldsymbol{x}_i^\top \boldsymbol{\beta}) \right\} = 0.$$

They used the same approach to robustify the likelihood equations for the unknown parameters, $\boldsymbol{\beta}$ and A, as

$$\sum_{i=1}^m \frac{\boldsymbol{x}_i^\top}{(A + D_i)^{1/2}} \psi_K \left(\frac{y_i - \boldsymbol{x}_i^\top \boldsymbol{\beta}}{(A + D_i)^{1/2}} \right) = 0,$$

$$\sum_{i=1}^m \left[\frac{1}{A + D_i} \left\{ \psi_K \left(\frac{y_i - \boldsymbol{x}_i^\top \boldsymbol{\beta}}{(A + D_i)^{1/2}} \right) \right\}^2 - \frac{1}{A + D_i} \right] = 0,$$

which can be solved by a Newton–Raphson algorithm. Note that the above equation reduces to the standard likelihood equation when $K \to \infty$.

Furthermore, Sugasawa (2020) proposed the use of density power divergence to obtain robust empirical Bayes estimators. A key consideration is that the best predictor of θ_i admits the following expression:

$$\widetilde{\theta}_i \equiv y_i - \frac{D_i}{A + D_i}(y_i - \boldsymbol{x}_i^\top \boldsymbol{\beta}) = y_i + D_i \frac{\partial}{\partial y_i} \log f(y_i; \boldsymbol{\beta}, A),$$

where $f(y_i; \boldsymbol{\beta}, A)$ is the density function of $y_i \sim \mathrm{N}(x_i^\top \boldsymbol{\beta}, A + D_i)$. The above expression is known as 'Tweedie's formula'. Since the unknown parameters $\boldsymbol{\beta}$ and A are estimated via the marginal density $f(y_i; \boldsymbol{\beta}, A)$, the form of $f(y_i; \boldsymbol{\beta}, A)$ is the key to the empirical Bayes estimator of θ_i. Thus, a simple idea to robustify the empirical Bayes estimator is to replace the marginal density with a robust alternative. Specifically, the robust likelihood function given by density power divergence applied to the Fay–Herriot model is given by

$$L_\alpha(y_i;\boldsymbol{\beta},A)=\frac{1}{\alpha}\phi(y_i;\boldsymbol{x}_i^\top\boldsymbol{\beta},A+D_i)^\alpha-\{2\pi(A+D_i)\}^{-\alpha/2}(1+\alpha)^{-3/2}.$$

Note that it holds that $\lim_{\alpha\to 0}\{L_\alpha(y_i;\boldsymbol{\beta},A)-(1/\alpha-1)\}\to\log f(y_i;\boldsymbol{\beta},A)$, so that the $L_\alpha(y_i;\boldsymbol{\beta},A)$ is a natural generalization of the marginal density. Then, we can define the robust Bayes estimator as

$$\begin{aligned}\widetilde{\theta}_i^{\mathrm{RD}}&\equiv y_i+D_i\frac{\partial}{\partial y_i}L_\alpha(y_i;\boldsymbol{\beta},A)\\&=y_i-\frac{D_i}{A+D_i}(y_i-\boldsymbol{x}_i^\top\boldsymbol{\beta})\phi(y_i;\boldsymbol{x}_i^\top\boldsymbol{\beta},A+D_i)^\alpha.\end{aligned}\tag{6.6}$$

Note that $\alpha>0$ is a tuning parameter controlling the degree of robustness (larger α generally leads to stronger robustness), and $\widetilde{\theta}_i^{\mathrm{RD}}$ reduces to the standard estimator $\widetilde{\theta}_i$ as $\alpha\to 0$. A notable property of $\widetilde{\theta}_i^{\mathrm{RD}}$ is that the second term in (6.6) converges to 0 as $|y_i-\boldsymbol{x}_i^\top\boldsymbol{\beta}|\to\infty$ under fixed parameter values of $\boldsymbol{\beta}$ and A. This means that if the auxiliary information $\boldsymbol{x}_i$ is not useful for y_i in some areas (i.e., the residual $|y_i-\boldsymbol{x}_i^\top\boldsymbol{\beta}|$ is large), the direct estimator is not much shrunken toward $\boldsymbol{x}_i^\top\boldsymbol{\beta}$ to prevent over-shrinkage. On the other hand, Sugasawa (2021) also showed that the two existing robust estimators by Ghosh et al. (2008) and Sinha and Rao (2009) do not have the property.

The MSE of $\widetilde{\theta}_i^{\mathrm{RD}}$ is $\mathrm{E}[(\widetilde{\theta}_i^{\mathrm{RD}}-\theta_i)^2]=g_{1i}(A)+g_{2i}(A,\alpha)$, where $g_{1i}(A)=AD_i/(A+D_i)$ and

$$g_{2i}(A,\alpha)=\frac{D_i^2}{(A+D_i)}\left\{\frac{V_i^{2\alpha}}{(2\alpha+1)^{3/2}}-\frac{2V_i^\alpha}{(\alpha+1)^{3/2}}+1\right\}.$$

Note that $g_{1i}(A)$ corresponds to the MSE of the non-robust Bayes estimator $\widetilde{\theta}_i$, so that $g_{2i}(A)$ is the inflation term due to the use of $\alpha>0$. Thus the choice of α can be done by specifying the tolerance percentage of inflation MSE.

Based on the robust likelihood function, one can obtain robust estimators of $\boldsymbol{\beta}$ and A as the maximizer of $\sum_{i=1}^m L_\alpha(y_i;\boldsymbol{\beta},A)$. The induced estimating equations are given by

$$\begin{aligned}\frac{\partial L_\alpha}{\partial\boldsymbol{\beta}}&=\sum_{i=1}^m\frac{\boldsymbol{x}_i\phi\left(y_i;\boldsymbol{x}_i^\top\boldsymbol{\beta},A\right)^\alpha\left(y_i-\boldsymbol{x}_i^\top\boldsymbol{\beta}\right)}{A+D_i}=0,\\2\frac{\partial L_\alpha}{\partial A}&=\sum_{i=1}^m\left\{\frac{\phi\left(y_i;\boldsymbol{x}_i^\top\boldsymbol{\beta},A\right)^\alpha}{(A+D_i)^2}\left\{\left(y_i-\boldsymbol{x}_i^\top\boldsymbol{\beta}\right)^2-(A+D_i)\right\}+\frac{\alpha V_i^\alpha}{(\alpha+1)^{3/2}(A+D_i)}\right\}=0,\end{aligned}$$

which can be solved by a Newton–Raphson algorithm.

Example 6.4 (*Simulation study*) The performance of robust small area estimators is compared through simulation studies in Sugasawa (2021). Consider the Fay–Herriot model:

$$y_i = \theta_i + \varepsilon_i, \quad \theta_i = \beta_0 + \beta_1 x_i + A^{1/2} u_i, \quad i = 1, \ldots, m,$$

where $m = 30$, $\beta_0 = 0$, $\beta_1 = 2$, $A = 0.5$, $\varepsilon_i \sim N(0, D_i)$ and $x_i \sim U(0, 1)$. Regarding the setting of D_i, m areas are divided into five groups containing an equal number of areas and set the same value of D_i within each group. The D_i pattern of the groups is (0.2, 0.4, 0.6, 0.8, 1.0). The following distribution is adopted for u_i:

$$u_i \sim (1 - \xi)\mathrm{N}(0, 1) + \xi N\left(0, 10^2\right),$$

where ξ determines the degree of misspecification of the assumed distribution. Two scenarios, (I) $\xi = 0$ and (II) $\xi = 0.15$, are considered. Note that in the latter scenario, some observations have very large residuals and auxiliary information x_i would not be useful for such outlying observations. To estimate θ_i, the standard empirical Bayes (EB) estimator; the robust estimators by Sinha and Rao (2009) (REB) and Ghosh et al. (2008) (GEB); and the robust estimator by Sugasawa (2021) (DEB). To determine K in GEB and α in DEB, 5% MSE inflation is adopted, $K = 1.345$ (a widely used value in Huber's ψ-function) is adopted in REB. Based on 20000 replications, MSE of the estimators of θ_i is computed and then they are aggregated. The obtained values are 1.45 (EB), 1.48 (REB), 1.64 (GEB), and 1.51 (DEB) under scenario (I), and 2.67 (EB), 2.55 (REB), 2.59 (GEB), and 2.21 (DEB) under scenario (II). Since the Fay–Herriot model is the true model, it is reasonable that EB performs best. However, once the distribution is misspecified in scenario (II), the performance of EB is not satisfactory. Although the two robust methods (REB and GEB) improve the estimation accuracy of EB in this scenario, DEB provides much better accuracy than the other robust methods. See Sugasawa (2021) for more detailed results.

There are some works on robust estimation of the Fay–Herriot model using heavy-tailed distributions. For example, Datta and Lahiri (1995) replaced the normal distribution for the random effects with the Cauchy distribution to capture outlying areas, and discussed robustness properties of the resulting estimator. Furthermore, Ghosh et al. (2018) studied the use of the t-distribution for the random effects.

References

Battese G, Harter R, Fuller W (1988) An error-components model for prediction of county crop areas using survey and satellite data. J Am Stat Assoc 83:28–36

Chambers R, Chandra H, Salvati N, Tzavidis N (2014) Outliner robust small area estimation. J Roy Stat Soc B 76:47–69

Chatterjee S, Lahiri P, Li H (2008) Parametric bootstrap approximation to the distribution of EBLUP and related predictions intervals in linear mixed models. Ann Stat 36:1221–1245

Datta GS, Lahiri P (1995) Robust hierarchical Bayes estimation of small area characteristics in the presence of covariates and outliers. J Multivar Anal 54:310–328

Dongmo-Jiongo V, Haziza D, Duchesne P (2013) Controlling the bias of robust small area estimators. Biometrika 100:843–858

Efron B, Morris CN (1975) Data analysis using Stein's estimator and its generalizations. J Am Stat Assoc 70:311–319

Ghosh M, Myung J, Moura FAS (2018) Robust Bayesian small area estimation. Surv Methodol 44:001-X

Hirose M (2017) Non-area-specific adjustment factor for second-order efficient empirical Bayes confidence interval. Comput Stat Data Anal 116:67–78

Hirose M (2019) A class of general adjusted maximum likelihood methods for desirable mean squared error estimation of EBLUP under the Fay-Herriot small area model. J Stat Plan Inference 199:302–310

Hirose M, Lahiri P (2018) Estimating variance of random effects to solve multiple problems simultaneously. Ann Stat 46:1721–1741

Jiang J, Nguyen T, Rao JS (2011) Best predictive small area estimation. J Am Stat Assoc 106:732–745

Li H, Lahiri P (2010) An adjusted maximum likelihood method for solving small area estimation problems. J Multivar Anal 101:882–892

Sinha SK, Rao JNK (2009) Robust small area estimation. Can J Stat 37:381–399

Sugasawa S (2021) Robust empirical Bayes small area estimation with density power divergence. Biometrika 107:467–480

Sugasawa S, Kawakubo Y, Datta GS (2019) Observed best selective prediction in small area estimation. J Multivar Anal 173:383–392

Yoshimori M, Lahiri P (2014) A new adjusted maximum likelihood method for the Fay-Herriot small area model. J Multivar Anal 124:281–294

Yoshimori M, Lahiri P (2014) A second-order efficient empirical Bayes confidence interval. Ann Stat 42:1233–1261

Chapter 7
Small Area Models for Non-normal Response Variables

As introduced in Chap. 4, the basic small area models are based on normality assumption for the response variables. However, we often need to handle non-normal response variables in practice. In this chapter, we review some techniques for small area estimation under non-normal data.

7.1 Generalized Linear Mixed Models

Generalized linear mixed models (GLMM) would be the most famous mixed-effects models to handle various types of response variables. Let y_{ij} be a response variable of the jth unit in the ith area, where $j = 1, \ldots, n_i$ and $i = 1, \ldots, m$. The GLMM for unit-level data assumes that y_{ij} given random effect v_i are conditionally independent and the conditional distribution belongs to the following exponential family distributions:

$$f(y_{ij}|v_i) = \exp\left\{\frac{\theta_{ij} y_{ij} - \psi(\theta_{ij})}{a_{ij}(\phi)} + c(y_{ij}, \phi)\right\}, \qquad \theta_{ij} = \boldsymbol{x}_{ij}^\top \boldsymbol{\beta} + v_i, \tag{7.1}$$

for $j = 1, \ldots, n_i$ and $i = 1, \ldots, m$, where $\psi(\cdot)$, $a_{ij}(\cdot)$ and $c(\cdot, \cdot)$ are known functions, $\boldsymbol{x}_{ij}$ is a vector of covariates, and ϕ is a dispersion parameter which may or may not be known. The quantity θ_{ij} is associated with the conditional mean, namely, $\mathrm{E}[y_{ij}|v_i] = \psi'(\theta_{ij})$ for a so-called canonical link function $\psi'(\cdot)$. Furthermore, it is also assumed that $v_i \sim \mathrm{N}(0, \tau^2)$. The nested error regression model given in Sect. 4.2 is the case with $\psi(x) = x^2/2$, $a_{ij}(\phi) = \phi$ and $c(y_{ij}, \phi) = -y_{ij}^2/2\phi$ in the model (7.1). Note that, in this case, the dispersion parameter $\phi = \sigma^2$ is unknown.

S. Sugasawa and T. Kubokawa, *Mixed-Effects Models and Small Area Estimation*, JSS Research Series in Statistics, https://doi.org/10.1007/978-981-19-9486-9_7

Under the model (7.1), the marginal distribution of $\boldsymbol{y}$ is

$$f(\boldsymbol{y}; \boldsymbol{\beta}, \phi, \tau^2) = \prod_{i=1}^{m} \int \exp\left\{ \frac{\theta_{ij}(v_i)y_{ij} - \psi(\theta_{ij}(v_i))}{a_{ij}(\phi)} + c(y_{ij}, \phi) \right\} \phi(v_i; 0.\tau^2) dv_i.$$

The unknown parameters can be estimated by maximizing the marginal distribution. However, in general, the above marginal distribution cannot be obtained in an analytical form except for the normal distribution for $y_{ij}|v_i$. For comprehensive theory of the parameter estimation of GLMM, see Jiang and Nguyen (2007). Ghosh et al. (1998) discussed the use of generalized linear mixed models in small area estimation and provided hierarchical Bayesian methods for fitting the model.

In what follows, we consider details of the prediction method using (7.1) under binary responses, as discussed in Jiang and Lahiri (2001). The unit-level model for binary data is given by

$$\mathrm{P}(y_{ij} = 1|p_{ij}) = p_{ij}, \quad \log\left(\frac{p_{ij}}{1 - p_{ij}}\right) = \boldsymbol{x}_{ij}^\top \boldsymbol{\beta} + v_i,$$

where $v_i \sim \mathrm{N}(0, \tau^2)$ is a random effect. The above model corresponds to the case with $\psi(x) = \log(1 + e^x)$ and $a_{ij}(\phi) = 1$ in the model (7.1). A typical purpose is to predict the true area proportion given by $\bar{p}_i = N_i^{-1} \sum_{j=1}^{N_i} y_{ij}$, where y_{ij} is observed only for $j = 1, \ldots, n_i (< N_i)$ and $\boldsymbol{x}_{ij}$ is observed for all N_i samples. The best predictor of the random effect v_i under squared loss is the conditional expectation of v_i given $\boldsymbol{y}_i = (y_{i1}, \ldots, y_{in_i})$, which is expressed as

$$\widetilde{v}_i(\boldsymbol{\beta}, \tau^2) \equiv \mathrm{E}[v_i|\boldsymbol{y}_i] = \frac{\int v_i \ell_i(\boldsymbol{y}_i; v_i, \beta, \tau^2)\phi(v_i; 0, \tau^2) dv_i}{\int \ell_i(\boldsymbol{y}_i; v_i, \beta, \tau^2)\phi(v_i; 0, \tau^2) dv_i} = \tau \frac{E_z[z\ell_i(\boldsymbol{y}_i; \tau z, \beta, \tau^2)]}{E_z[\ell_i(\boldsymbol{y}_i; \tau z, \beta, \tau^2)]},$$

where $z \sim \mathrm{N}(0, 1)$, E_z denotes the expectation with respect to z, and

$$\log \ell_i(\boldsymbol{y}_i; v_i, \beta, \tau^2) = v_i \sum_{j=1}^{n_i} y_{ij} - \sum_{j=1}^{n_i} \log\{1 + \exp(\boldsymbol{x}_{ij}^\top \boldsymbol{\beta} + v_i)\}.$$

Although the conditional expectation does not admit an analytical expression, it can be numerically computed by Monte Carlo integration, that is, the best predictor can be approximated as

$$\mathrm{E}[v_i|\boldsymbol{y}_i] \approx \tau \frac{R^{-1} \sum_{r=1}^{R} z^{(r)} \ell_i(\boldsymbol{y}_i; \tau z^{(r)}, \beta, \tau^2)}{R^{-1} \sum_{r=1}^{R} \ell_i(\boldsymbol{y}_i; \tau z^{(r)}, \beta, \tau^2)},$$

where $z^{(r)}$ is an independent random sample from N(0, 1) and R is the Monte Carlo samples. Under $R \to \infty$, the above approximation converges to the exact conditional expectation. Let $\widehat{\boldsymbol{\beta}}$ and $\widehat{\tau}^2$ be the maximum likelihood estimators obtained by maximizing the marginal log-likelihood:

$$\sum_{i=1}^{m} \log\left\{\int \ell_i(\boldsymbol{y}_i; v_i, \beta, \tau^2)\phi(v_i; 0, \tau^2)dv_i\right\}.$$

Hence, the unobserved y_{ij} can be predicted as $\widehat{p}_{ij} = \text{logit}(\boldsymbol{x}_{ij}^\top\widehat{\boldsymbol{\beta}} + \widetilde{v}_i(\widehat{\boldsymbol{\beta}}, \widehat{\tau}^2))$ for $j = n_i + 1, \ldots, N_i$. Finally, the population mean $\bar{p}_i$ is estimated as

$$\widehat{\bar{p}}_i = \frac{1}{N_i}\left\{\sum_{j=1}^{n_i} y_{ij} + \sum_{j=n_i+1}^{N_i} \widehat{p}_{ij}\right\}, \qquad i = 1, \ldots, m.$$

7.2 Natural Exponential Families with Conjugate Priors

One of the main drawbacks of generalized linear mixed models (7.1) is the intractability of conditional distributions of random effects as well as marginal likelihood functions. As alternative models for area-level non-normal data, Ghosh and Maiti (2004) introduced models based on the natural exponential families with conjugate priors. Let $y_1, \ldots, y_m$ be mutually independent random variables where the conditional distribution of y_i given θ_i and the prior distribution of θ_i belong to the following natural exponential families:

$$\begin{aligned} f(y_i|\theta_i) &= \exp[n_i\{\theta_i y_i - \psi(\theta_i)\} + c(y_i, n_i)], \\ \pi(\theta_i) &= \exp[\nu\{m_i\theta_i - \psi(\theta_i)\}]C(\nu, m_i), \end{aligned} \tag{7.2}$$

where n_i is a known scalar and ν is an unknown scalar. Here, $c(\cdot, \cdot)$ and $C(\cdot, \cdot)$ are normalizing constants and $\psi(\cdot)$ is a link function. Moreover, $m_i = \psi'(\boldsymbol{x}_i^\top\boldsymbol{\beta})$, where $\boldsymbol{x}_i$ and $\boldsymbol{\beta}$ are vectors of covariates and unknown regression coefficients, respectively, and $\psi'(\cdot)$ is the first-order derivative of $\psi(\cdot)$. Usually, n_i is related to sample size in the ith area, so that n_i is not so large in practice. The function $f(y_i|\theta_i)$ is the regular one-parameter exponential family and the function $\pi(\theta_i)$ is the conjugate prior distribution. Note that $\mu_i \equiv \text{E}[y_i|\theta_i] = \psi'(\theta_i)$ which is the true area mean, and $\text{E}[\mu_i] = m_i$. Therefore, y_i is regarded as a crude estimator of μ_i and m_i is the prior mean of μ_i. Ghosh and Maiti (2004) focused on exponential family with quadratic variance functions, a slightly narrower class of exponential family distributions, where the conditional variance is $\text{Var}(y_i|\theta_i) = V(\mu_i)/n_i$ for a quadratic function of $V(\cdot)$. Such a family includes representative models that are often used in practice. For example, when $\nu = A^{-1}$ and $n_i = D_i^{-1}$ and $\psi(x) = x^2/2$, the model (7.2) reduces to the Fay–Herriot model (4.1). Also, when $\psi(x) = \exp(x)$ and $\psi(x) = \log(1 + \exp(x))$, the model (7.2) reduces to Poisson-gamma models (Clayton and Kalder 1987) and binomial-beta models (Williams 19075), respectively.

Owing to the conjugacy, the marginal likelihood can be obtained in a closed form, which enables us to get the maximum likelihood estimators or moment-type estimators for the unknown parameters, $\boldsymbol{\beta}$ and ν. Moreover, the conditional distribution of θ_i given y_i can be obtained as

$$\pi(\theta_i | y_i) \propto \exp\{\theta_i(n_i y_i + \nu m_i) - (n_i + \nu)\psi(\theta_i)\},$$

so that the conditional expectation of μ_i is given by

$$\widetilde{\mu}_i(\boldsymbol{\beta}, \nu) \equiv \mathrm{E}[\mu_i | y_i] = \frac{n_i y_i + \nu m_i}{n_i + \nu}.$$

It is observed that $\widetilde{\mu}_i$ is a weighted average of the direct estimator y_i and the estimator of prior mean (regression estimator) m_i. A typical way to estimate $\boldsymbol{\beta}$ and A is to maximize the marginal log-likelihood, namely

$$(\widehat{\boldsymbol{\beta}}, \widehat{\nu}) = \mathrm{argmax}_{\boldsymbol{\beta}, \nu} \sum_{i=1}^{m} \{\log C(\nu, m_i) - \log C(n_i + \nu, \widetilde{\mu}_i(\boldsymbol{\beta}, \nu))\}.$$

On the other hand, Ghosh and Maiti (2004) employed the optimal estimating equation with moment conditions. Given the parameter estimates, the empirical Bayes estimator of the true mean μ_i is $\widetilde{\mu}_i(\widehat{\boldsymbol{\beta}}, \widehat{\nu})$.

For measuring the uncertainty of the empirical Bayes estimator, Ghosh and Maiti (2004) derived the second-order unbiased MSE estimator, and Ghosh and Maiti (2008) constructed empirical Bayes confidence intervals.

7.3 Unmatched Sampling and Linking Models

Suppose that the response variable y_i is continuous but normality assumption seems not suitable. Such examples include continuous positive values or proportions. In this case, You and Rao (2002) proposed the following extension of the standard Fay–Herriot model (4.1):

$$y_i = \theta_i + \varepsilon_i, \quad h(\theta_i) = \boldsymbol{x}_i^\top \boldsymbol{\beta} + v_i, \quad i = 1, \ldots, m, \tag{7.3}$$

where $\varepsilon_i \sim \mathrm{N}(0, D_i)$, $v_i \sim \mathrm{N}(0, A)$, and $h(\cdot)$ is a known link function. Note that the use of the identity link $h(x) = x$ reduces to the standard Fay–Herriot model. A typical example for positive valued y_i is $h(x) = \log x$. The model (7.3) is called unmatched sampling and linking models. You and Rao (2002) proposed a hierarchical Bayes approach and Sugasawa et al. (2018) developed an empirical Bayes approach for fitting the model (7.3).

In what follows, we explain the empirical Bayes approach to obtain the empirical best predictor of θ_i. Under the model (7.3), the joint density of y_i and θ_i can be expressed as

$$f\left(y_i, \theta_i\right) = \frac{h'\left(\theta_i\right)}{2\pi\left(AD_i\right)^{1/2}} \exp\left\{-\frac{\left(y_i-\theta_i\right)^2}{2D_i} - \frac{\left(h\left(\theta_i\right)-\boldsymbol{x}_i^\top\boldsymbol{\beta}\right)^2}{2A}\right\},$$

where h' is the first-order derivative of h. Then, the marginal density of y_i is

$$f\left(y_i\right) = \frac{1}{2\pi\left(AD_i\right)^{1/2}} \int_{\Theta} h'\left(\theta_i\right) \exp\left\{-\frac{\left(y_i-\theta_i\right)^2}{2D_i} - \frac{\left(h\left(\theta_i\right)-\boldsymbol{x}_i^\top\boldsymbol{\beta}\right)^2}{2A}\right\} d\theta_i.$$

The best predictor or Bayes estimator of θ_i under squared error loss is given by the conditional expectation $\mathrm{E}[\theta_i|y_i]$, which has the following expression:

$$\mathrm{E}\left(\theta_i \mid y_i\right) = f\left(y_i\right)^{-1} \int_{\Theta} \theta_i f\left(y_i, \theta_i\right) d\theta_i.$$

By changing the variable $h(\theta_i)$ to $z = A^{-1/2}\{h(\theta_i - \boldsymbol{x}_i^\top\boldsymbol{\beta})\}$ in the two integrals appeared in $\mathrm{E}[\theta_i|y_i]$, we obtain an alternative expression:

$$\widetilde{\theta}_i(y_i; \boldsymbol{\beta}, A) \equiv \mathrm{E}\left(\theta_i \mid y_i\right) = \frac{\mathrm{E}_z\left[\theta_i^* \exp\left\{-\left(2D_i\right)^{-1}\left(y_i-\theta_i^*\right)^2\right\}\right]}{\mathrm{E}_z\left[\exp\left\{-\left(2D_i\right)^{-1}\left(y_i-\theta_i^*\right)^2\right\}\right]},$$

where $\theta_i^* = h^{-1}(\sqrt{A}z + \boldsymbol{x}_i^\top\boldsymbol{\beta})$ and E_z denotes the expectation with respect to $z \sim$ N(0, 1). Although the closed-form of $\widetilde{\theta}_i(y_i; \boldsymbol{\beta}, A)$ cannot be obtained in general, it can be easily computed via Monte Carlo integration by generating a large number of z from N(0, 1) or the deterministic approximation using Gauss–Hermite quadrature.

For maximizing the marginal likelihood of the unknown parameters, an Expectation–Maximization (EM) algorithm can be used. Define the 'complete' log-likelihood as

$$L^c(\boldsymbol{\beta}, A) = C - \frac{m}{2}\log A - \frac{1}{2}\sum_{i=1}^m \frac{(y_i-\theta_i)}{D_i} - \frac{1}{2A}\sum_{i=1}^m \{h(\theta_i) - \boldsymbol{x}_i^\top\boldsymbol{\beta}\}^2,$$

where C is a generic constant that does not depend on the parameters. By taking expectations of $L^C(\boldsymbol{\beta}, A)$ with respect to the conditional distribution of $\theta_i|y_i$, we obtain the following objective function:

$$Q(\boldsymbol{\beta}, A|\boldsymbol{\beta}^{(r)}, A^{(r)}) = -\frac{m}{2}\log A - \frac{1}{2A}\sum_{i=1}^m \mathrm{E}^{(r)}\left[\{h(\theta_i) - \boldsymbol{x}_i^\top\boldsymbol{\beta}\}^2\right],$$

where $\mathrm{E}^{(r)}$ denotes the expectation with $\boldsymbol{\beta} = \boldsymbol{\beta}^{(r)}$ and $A = A^{(r)}$. Hence, the updating steps are given by

$$\boldsymbol{\beta}^{(r+1)} = \left(\sum_{i=1}^{m} \boldsymbol{x}_i \boldsymbol{x}_i^\top\right)^{-1} \sum_{i=1}^{m} \boldsymbol{x}_i \mathrm{E}^{(r)}\left[h\left(\theta_i\right)\right],$$

$$A^{(r+1)} = \frac{1}{m} \sum_{i=1}^{m} \mathrm{E}^{(r)}\left[\left\{h\left(\theta_i\right) - \boldsymbol{x}_i^\top \boldsymbol{\beta}^{(r+1)}\right\}^2\right].$$

It is noted that the expectation $\mathrm{E}^{(r)}[g(\theta_i)]$ for some function g has the expression

$$\mathrm{E}^{(r)}\left[g\left(\theta_i\right)\right] = \frac{\mathrm{E}_z\left[g\left(\theta_i^{(r)}(z)\right)\exp\left\{-\left(2D_i\right)^{-1}\left(y_i - \theta_i^{(r)}(z)\right)^2\right\}\right]}{\mathrm{E}_z\left[\exp\left\{-\left(2D_i\right)^{-1}\left(y_i - \theta_i^{(r)}(z)\right)^2\right\}\right]},$$

where $\theta_i^{(r)}(z) = h^{-1}(\sqrt{A^{(r)}}z + \boldsymbol{x}_i^\top \boldsymbol{\beta}^{(r)})$ and $z \sim \mathrm{N}(0, 1)$. Therefore, the integral can be computed by Monte Carlo integration or Gauss–Hermite quadrature. Starting with some initial values of $\boldsymbol{\beta}$ and A, the EM algorithm repeats the above updates until convergence. Finally, the empirical Bayes estimator of θ_i is $\widetilde{\theta}_i(y_i; \widehat{\boldsymbol{\beta}}, \widehat{A})$.

Although the model (7.3) assumes a known link function $h(\cdot)$, it is possible to estimate the link function based on the data. First, rewrite the model (7.3) as

$$y_i = \theta_i + \varepsilon_i, \quad \theta_i = L(\boldsymbol{x}_i^\top \boldsymbol{\beta} + v_i), \quad i = 1, \ldots, m,$$

where $L(\cdot)$ is an unknown function. Sugasawa and Kubokawa (2019) proposed estimating the unknown link function via P-splines given by

$$L\left(x; \boldsymbol{\gamma}\right) = \gamma_{10} + \gamma_{11}x + \cdots + \gamma_{1q}x^q + \sum_{j=1}^{K} \gamma_{2j}\left(x - \kappa_j\right)_+^q,$$

where q is the degree of the spline, $(x)_+^q$ denotes the function $x^q I_{x>0}$, $\kappa_1 < \cdots < \kappa_K$ is a set of fixed knots, and $\boldsymbol{\gamma}$ is a coefficient vector. Then, the model can be rewritten as

$$y_i | u_i, \boldsymbol{\gamma}_2 \sim \mathrm{N}(\boldsymbol{z}_1(u_i)^\top \boldsymbol{\gamma}_1 + \boldsymbol{z}_2(u_i; \delta)^\top \boldsymbol{\gamma}_2, D_i),$$
$$u_i \sim \mathrm{N}(\boldsymbol{x}_i^\top \boldsymbol{\beta}, A), \quad \boldsymbol{\gamma}_2 \sim \mathrm{N}(0, \lambda \boldsymbol{I}_K),$$

where $\boldsymbol{z}_1(u_i) = (1, u_i, \ldots, u_i^q)^\top$, $\boldsymbol{\gamma}_1 = (\gamma_{10}, \ldots, \gamma_{1q})^\top$ and $\boldsymbol{z}_2(u_i; \delta) = ((u_i - \kappa_1)_+^q, \ldots, (u_i - \kappa_K)_+^q)^\top$. Here, $\theta_i = \boldsymbol{z}_1(u_i)^\top \boldsymbol{\gamma}_1 + \boldsymbol{z}_2(u_i; \delta)^\top \boldsymbol{\gamma}_2$ is the small area parameter. Sugasawa and Kubokawa (2019) put prior distributions on the unknown parameters and developed a hierarchical Bayesian method for estimating θ_i.

7.4 Models with Data Transformation

7.4.1 *Area-Level Models for Positive Values*

In practice, we often deal with positive response variables such as income, for which the normality assumption used in the Fay–Herriot model may not be reasonable. To address this issue, the log-transformation is widely adopted due to simplicity. Slud and Maiti (2006) investigated theoretical properties of the log-transformed Fay–Herriot model, given by

$$\log y_i = \theta_i + \varepsilon_i, \quad \theta_i = \boldsymbol{x}_i^\top \boldsymbol{\beta} + v_i, \quad i = 1, \dots, m,$$

where $\varepsilon_i \sim \mathrm{N}(0, D_i)$, $v_i \sim \mathrm{N}(0, A)$, and the target parameter of interest is $\exp(\theta_i)$. Note the conditional distribution of θ_i given y_i is $\mathrm{N}(\widetilde{\theta}_i, s_i^2)$, where

$$\widetilde{\theta}_i = \boldsymbol{x}_i^\top \boldsymbol{\beta} + \gamma_i(\log y_i - \boldsymbol{x}_i^\top \boldsymbol{\beta}), \qquad \gamma_i = A/(A + D_i),$$

and $s_i^2 = AD_i/(A + D_i)$. Then, it holds that

$$\mathrm{E}[\exp(\theta_i)|y_i] = \exp\left\{\widetilde{\theta}_i + \frac{1}{2}s_i^2\right\} = \exp\left\{\boldsymbol{x}_i^\top \boldsymbol{\beta} + \gamma_i(\log y_i - \boldsymbol{x}_i^\top \boldsymbol{\beta}) + \frac{AD_i}{2(A + D_i)}\right\}.$$

Although the log-transformation is analytically tractable, a potential drawback is that the log-transformation is not necessarily reasonable. Rather, it would be more preferable to use a parametric family of transformations and estimate the transformation parameter based on the data.

Sugasawa and Kubokawa (2015) introduced a parametric transformed Fay–Herriot model:

$$H(y_i; \lambda) = \boldsymbol{x}_i^\top \boldsymbol{\beta} + v_i + \varepsilon_i, \quad i = 1, \dots, m, \tag{7.4}$$

where $H(y_i; \lambda)$ is a parametric transformation and λ is an unknown transformation parameter. The log-likelihood function (without irrelevant constant terms) of the unknown parameters is obtained as

$$L(\boldsymbol{\beta}, A, \lambda) = -\sum_{i=1}^m \log(A + D_i) - \sum_{i=1}^m \frac{\left\{H(y_i; \lambda) - \boldsymbol{x}_i^\top \boldsymbol{\beta}\right\}^2}{A + D_i} + 2\sum_{i=1}^m \log H'(y_i; \lambda),$$

where $H'(x; \lambda) = \partial H(x; \lambda)/\partial x$ is the partial derivative of the transformation function. Sugasawa and Kubokawa (2015) established asymptotic properties of the maximum likelihood estimator under suitable conditions for the transformation function. As an example, Sugasawa and Kubokawa (2015) suggested the dual power transformation (Yang 2006) defined as

$$H(x; \lambda) = \begin{cases} (2\lambda)^{-1} \left(x^{\lambda} - x^{-\lambda}\right) & \lambda > 0 \\ \log x & \lambda = 0, \end{cases} \tag{7.5}$$

which includes the log-transformation as a special case. Hence, the dual power transformation can be a useful alternative to the log-transformation.

Sugasawa and Kubokawa (2017) derived the best predictor of $\eta_i \equiv H^{-1}(\theta_i; \lambda)$, where $H^{-1}(x; \lambda)$ is the inverse function of $H(x; \lambda)$ as a function of x. Since the conditional distribution of θ_i given y_i under the model (7.4) is $\mathrm{N}(\widetilde{\theta}_i, s_i^2)$, where

$$\widetilde{\theta}_i = \boldsymbol{x}_i^\top \boldsymbol{\beta} + \gamma_i \left\{H(y_i; \lambda) - \boldsymbol{x}_i^\top \boldsymbol{\beta}\right\}, \quad \gamma_i = A/(A + D_i).$$

Then, the best predictor of η_i is

$$\widetilde{\eta}(y_i; \boldsymbol{\beta}, A, \lambda) \equiv \mathrm{E}[\eta_i | y_i] = \int_{-\infty}^{\infty} H^{-1}(t; \lambda)\phi(t; \widetilde{\theta}_i, s_i^2)dt.$$

Given the parameter estimates, the empirical best predictor is obtained as $\widetilde{\eta}(y_i; \widehat{\boldsymbol{\beta}}, \widehat{A}, \widehat{\lambda})$. However, for a general transformation function, the above expression cannot be obtained in an analytical form, but it can be approximated by Monte Carlo integration. As shown in Sugasawa and Kubokawa (2017), it is possible to derive the second-order unbiased mean squared error estimators via the parametric bootstrap.

Example 7.1 (*Estimating household expenditure*) Sugasawa and Kubokawa (2017) fitted the model (7.4) to survey data in Japan, and found that the estimated parameters in the dual power transformation are significantly away from 0, suggesting that the log-transformation may not be reasonable.

7.4.2 Area-Level Models for Proportions

Hirose et al. (2023) proposed a transformation model with arc-sin transformation. Let $y_1, \ldots, y_m$ be binomial observations, distributed as $y_i | p_i \sim \mathrm{Bin}(n_i, p_i)$ for $i = 1, \ldots, m$, where p_i is the area-wise proportion and n_i is the sample size in each area. The arc-sin transformation is given by $z_i = \sin^{-1}(2y_i - 1)$ with the corresponding parameters $\theta_i = \sin^{-1}(2p_i - 1)$. This transformation is known as 'variance stabilizing transformation' for estimating proportions. We consider the Fay–Herriot model for z_i, namely

$$z_i | \theta_i \sim \mathrm{N}(\theta, D_i), \quad \theta_i \sim \mathrm{N}(\boldsymbol{x}_i^\top \boldsymbol{\beta}, A), \quad i = 1, \ldots, m,$$

where $D_i = 1/4n_i$. The normality assumption of $z_i | \theta_i$ is based on the asymptotic normal approximation of $\mathrm{Bin}(n_i, p_i)$ under large n_i. Note that the parameter of interest is $p_i = \{\sin(\theta_i) + 1\}/2$.

From the standard theory of the Fay–Herriot model, it follows that

$$\theta_i|z_i \sim \mathrm{N}(\widetilde{\theta}_i, (1-B_i)/4n_i), \quad \widetilde{\theta}_i = z_i - B_i(z_i - x_i^\top \boldsymbol{\beta}),$$

where $B_i = 1/(1+4n_iA)$. Using the fact that $\mathrm{E}[\sin X] = \sin(a)\exp(-b/2)$ for $X \sim \mathrm{N}(a, b)$, the best predictor of p_i is given by

$$\begin{aligned}\widetilde{p}_i(\boldsymbol{\beta}, A) \equiv \mathrm{E}[p_i|z_i] &= \frac{1}{2}(1+\mathrm{E}[\sin\theta_i|z_i]) \\ &= \frac{1}{2}\left\{1+\sin(\widetilde{\theta}_i)\exp\left(-\frac{1-B_i}{8n_i}\right)\right\}. \qquad (7.6)\end{aligned}$$

By replacing $\boldsymbol{\beta}$ and A with their estimates $\widehat{\boldsymbol{\beta}}$ and $\widehat{A}$ using available methods in the Fay–Herriot model (e.g., residual maximum likelihood method), we obtain the empirical best predictor, $\widehat{p}_i = \widetilde{p}_i(\widehat{\boldsymbol{\beta}}, \widehat{A})$. For measuring uncertainty of $\widehat{p}_i$, Hirose et al. (2023) derived the second-order approximation of the MSE of $\widehat{p}_i$.

Example 7.2 (*Predicting the positive rate in PCR testing for each 47 prefectures in Japan*) Hirose et al. (2023) demonstrated the use of the arc-sin transformation model to the number of positive cases (y_i) among the number of people who have taken the PCR test (n_i) for $m = 47$ prefectures in Japan. The goal of the analysis was to estimate the prefecture-wise positive rate. Although the original sample size n_i is large, they tried a hypothetical setting with sample size $n_i^* = \lceil n_i \times 10^{-4} \rceil$, where $\lceil n \rceil$ indicates the smallest integer greater than or equal to the value of n. Consideration of such a situation is motivated by the stable estimation of the positive rate at the early stages of the pandemic. They compared the stability of several empirical Bayes estimates via the coefficient of variation defined as $\mathrm{CV}_i = \sqrt{\widehat{\mathrm{MSE}}_i}/\widehat{p}_i$, where $\widehat{\mathrm{MSE}}_i$ is the second-order unbiased MSE estimator of $\widehat{p}_i$, and the results suggest that the CV of the empirical best predictor (7.6) was much smaller than the other estimates. See Hirose et al. (2023) for more details.

7.4.3 Unit-Level Models and Estimating Finite Population Parameters

The same idea can be incorporated into the nested error regression model. Suppose that we are interested in the estimation of the general finite population parameters given by

$$\mu_i = \frac{1}{N_i}\sum_{j=1}^{N_i} T(Y_{ij}), \quad i = 1, \ldots, m,$$

where $T(\cdot)$ is a known function. For example, if we adopt $T(x) = I(x < z)$ for a fixed threading value z and Y_{ij} is a welfare measure, then μ_i can be interpreted as the poverty rate in the ith area.

To estimate μ_i, Molina and Rao (2010) adopted the transformed nested error regression model:

$$H(Y_{ij}) = \boldsymbol{x}_{ij}^{\top}\boldsymbol{\beta} + v_i + \varepsilon_{ij}, \quad j = 1, \dots, N_i, \quad i = 1, \dots, m, \tag{7.7}$$

where $v_i \sim \mathrm{N}(0, \tau^2)$, $\varepsilon_{ij} \sim \mathrm{N}(0, \sigma^2)$, and $H(\cdot)$ is a specified transformation function such as the logarithm function. We here assume that we observe y_{ij} only for $j = 1, \dots, n_i (< N_i)$ and $\boldsymbol{x}_{ij}$ for all the units. For notational convenience, we define $s_i = \{1, \dots, n_i\}$ and $r_i = \{n_i + 1, \dots, N_i\}$. Under the transformed model, the conditional distribution of $H(Y_{ij})$ for $j \in r_i$ given sampled units is given by

$$H(Y_{ij})|(y_{ij}, j \in s_i) \sim \mathrm{N}(\boldsymbol{x}_{ij}^{\top}\boldsymbol{\beta} + \widetilde{v}_i, \sigma^2), \quad j \in r_i,$$

where

$$\widetilde{v}_i = \frac{n_i\tau^2}{\sigma^2 + n_i\tau^2}\sum_{j=1}^{n_i}\left\{H(y_{ij}) - \boldsymbol{x}_{ij}^{\top}\boldsymbol{\beta}\right\}.$$

Then, the best predictor of μ_i is the conditional expectation

$$\mathrm{E}[\mu_i|y_{ij}, j \in s_i] = \frac{1}{N_i}\left\{\sum_{j\in s_i} T(y_{ij}) + \sum_{j\in r_i}\mathrm{E}[T(Y_{ij})|y_{ij}, j \in s_i]\right\}.$$

For general functions $T(\cdot)$ and $H(\cdot)$, the conditional expectation of $T(Y_{ij})$ cannot be obtained in an analytical form, but it can be computed via Monte Carlo integration by generating random samples of Y_{ij}, where Y_{ij} can be easily simulated via $H^{-1}(U_{ij})$ with U_{ij} generated from $\mathrm{N}(\boldsymbol{x}_{ij}^{\top}\boldsymbol{\beta} + \widetilde{v}_i, \sigma^2)$. The unknown parameters in the model (7.7) can be estimated via the existing methods for fitting the standard nested error regression models with transformed observations.

As mentioned in the previous subsection, the use of a specified transformation is subject to misspecification. To overcome the issue, Sugasawa and Kubokawa (2019) proposed the following parametric transformed nested error regression model:

$$H(y_{ij}; \lambda) = \boldsymbol{x}_{ij}^{\top}\boldsymbol{\beta} + v_i + \varepsilon_{ij}, \quad j = 1, \dots, N_i, \quad i = 1, \dots, m, \tag{7.8}$$

where $H(y_{ij}; \lambda)$ denotes transformed response variables. It should be noted that the method for estimating finite population parameter μ_i by Molina and Rao (2010) can be easily modified by replacing $H(\cdot)$ with $H(\cdot; \lambda)$. The log-likelihood function without irrelevant constants is given by

$$L(\boldsymbol{\beta}, \tau^2, \sigma^2, \lambda) = \sum_{i=1}^{m}\sum_{j=1}^{n_i} \log \frac{\partial}{\partial y_{ij}} H(y_{ij}; \lambda) - \frac{1}{2}\sum_{i=1}^{m} \log |\boldsymbol{\Sigma}_i|$$
$$- \frac{1}{2}\sum_{i=1}^{m} \left\{H(\boldsymbol{y}_i; \lambda) - \boldsymbol{X}_i\boldsymbol{\beta}\right\}^\top \boldsymbol{\Sigma}_i^{-1} \left\{H(\boldsymbol{y}_i; \lambda) - \boldsymbol{X}_i\boldsymbol{\beta}\right\}$$

where $\boldsymbol{y}_i = (y_{i1}, \ldots, y_{in_i})^\top$, $H(\boldsymbol{y}_i; \lambda) = (H(y_{i1}; \lambda), \ldots, H(y_{in_i}; \lambda))^\top$, $\boldsymbol{X}_i = (\boldsymbol{x}_{i1}, \ldots, \boldsymbol{x}_{in_i})^\top$ and $\boldsymbol{\Sigma} = \tau^2 \boldsymbol{J}_{n_i} + \sigma^2 \boldsymbol{I}_{n_i}$ with $\boldsymbol{J}_{n_i}$ being $n_i \times n_i$ matrix of 1's. Under given λ, the maximization with respect to the other parameters is equivalent to the maximum likelihood estimator of the nested error regression model with response variable $H(y_{ij}, \lambda)$, thereby the profile likelihood for λ can be easily computed. Sugasawa and Kubokawa (2019) adopted the golden section method for maximizing the profile likelihood of λ. Sugasawa and Kubokawa (2019) derived asymptotic properties of the estimator under some regularity conditions including restrictions for a parametric class of transformations.

There are several options for the parametric transformation $H_\lambda(\cdot)$. A representative function that includes the log-transformation is the dual power transformation (7.5), and its shifted version $H_{\lambda,c}(x) = \{(x+c)^\lambda - (x+c)^{-\lambda}\}/2\lambda$ with $c \in (\min(y_{ij}) + \varepsilon, \infty)$ with some small $\varepsilon > 0$ is also useful. Moreover, Jones and Pewey (2009) introduced the sinh-arcsinh transformation

$$H_{a,b}(x) = \sinh\left(b \sinh^{-1}(x) - a\right), \quad x \in (-\infty, \infty), a \in (-\infty, \infty), \quad b \in (0, \infty),$$

which can be applied for flexible modeling real-valued data. Regarding the choice of different parametric transformations, it would be useful to use information criteria of the form: $2\text{ML} + 2(p+q+2)$, where ML is the maximum log-likelihood, q is the number of transformation parameters, and $N = \sum_{i=1}^{m} n_i$.

For measuring uncertainty of the estimator of μ_i, we consider empirical Bayes confidence intervals of μ_i. A key to the derivation is the conditional distribution of μ_i given $\boldsymbol{y}_i$. Note that $\text{Cov}(H_\lambda(y_{ij}), H_\lambda(y_{ik}) \mid \boldsymbol{y}_i) = \text{Var}(v_i \mid \boldsymbol{y}_i) = s_i^2$ for $j \neq k$, where $s_i^2 = \sigma^2\tau^2/(\sigma^2 + n_i\tau^2)$. Then, it follows that

$$\left(H_\lambda\left(Y_{i,n_i+1}\right), \ldots, H_\lambda\left(Y_{iN_i}\right)\right)^\top \mid \boldsymbol{y}_i$$
$$\sim N\left(\left(\theta_{i,n_i+1}, \ldots, \theta_{iN_i}\right)^\top, s_i^2 \boldsymbol{1}_{N_i-n_i}\boldsymbol{1}_{N_i-n_i}^\top + \sigma^2 \boldsymbol{I}_{N_i-n_i}\right),$$

namely, each component has the expression

$$H_\lambda\left(Y_{ij}\right) \mid \boldsymbol{y}_i = \boldsymbol{x}_{ij}^\top\boldsymbol{\beta} + \widetilde{v}_i + s_i z_i + \sigma w_{ij}, \quad j = n_i + 1, \ldots, N_i,$$

where z_i and w_{ij} are mutually independent standard normal random variables. Then, the conditional distribution of μ_i is expressed as

$$\frac{1}{N_i}\left\{\sum_{j=1}^{n_i} T\left(y_{ij}\right)+\sum_{j=n_i+1}^{N_i} T\circ H_\lambda^{-1}\left(\boldsymbol{x}_{ij}^\top\boldsymbol{\beta}+\widetilde{v}_i+s_i z_i+\sigma w_{ij}\right)\right\} \tag{7.9}$$

which is a complex function of standard normal random variables z_i and w_{ij}. However, random samples from the distribution (7.9) can be easily simulated. We then define $Q_\alpha\left(y_i, \boldsymbol{\phi}_0\right)$ as the lower $100\alpha\%$ quantile point of the conditional distribution of μ_i with the true parameter $\boldsymbol{\phi}_0$ of $\boldsymbol{\phi}=(\boldsymbol{\beta},\sigma^2,\tau^2,\lambda)$, which satisfies $\mathrm{P}\left(\mu_i \le Q_\alpha\left(y_i,\boldsymbol{\phi}_0\right) \mid \boldsymbol{y}_i\right)=\alpha$. Hence, the interval of μ_i with nominal level $1-\alpha$ is obtained as $I_\alpha(\boldsymbol{\phi}_0)=\left(Q_{\alpha/2}\left(y_i,\boldsymbol{\phi}_0\right), Q_{1-\alpha/2}\left(y_i,\boldsymbol{\phi}_0\right)\right)$, which holds that $\mathrm{P}\left(\mu_i\in I_\alpha(\boldsymbol{\phi}_0)\right)=1-\alpha$. Since the interval $I_\alpha(\boldsymbol{\phi}_0)$ depends on the unknown parameter $\boldsymbol{\phi}_0$, the feasible version is obtained as $I_\alpha(\widehat{\boldsymbol{\phi}})$. It can be shown that $\mathrm{P}(\mu_i\in I_\alpha(\widehat{\boldsymbol{\phi}}))=1-\alpha+O(m^{-1})$, but the approximation error cannot be negligible when m is not large. The coverage accuracy can be improved by using the parametric bootstrap. To this end, we define the bootstrap estimator of the coverage probability of the plug-in interval $I_\alpha(\widehat{\boldsymbol{\phi}})$. Let Y_{ij}^* be the parametric bootstrap samples generated from the transformed nested error regression model with $\boldsymbol{\phi}=\widehat{\boldsymbol{\phi}}$, and $\boldsymbol{y}_i^*=(Y_{i1}^*,\ldots,Y_{in_i}^*)$. Moreover, let μ_i^* be the bootstrap version of μ_i based on Y_{ij}^*'s. Then, the parametric bootstrap estimator of the coverage probability is given by

$$\mathrm{CP}(\alpha)=E^*\left[I\left\{Q_{\alpha/2}\left(\boldsymbol{y}_i^*,\widehat{\boldsymbol{\phi}}^*\right)\le\mu_i^*\le Q_{1-\alpha/2}\left(\boldsymbol{y}_i^*,\widehat{\boldsymbol{\phi}}^*\right)\right\}\right],$$

where the expectation is taken with respect to the bootstrap samples. We define the calibrated nominal level a^* as the solution of the equation $\mathrm{CP}\left(a^*\right)=1-\alpha$, which can be solved, for example, by the bisection method. Then, the calibrated interval is given by

$$I_\alpha^C(\widehat{\boldsymbol{\phi}})=\left(Q_{\alpha^*/2}\left(\boldsymbol{y}_i,\widehat{\boldsymbol{\phi}}\right), Q_{1-\alpha^*/2}\left(\boldsymbol{y}_i,\widehat{\boldsymbol{\phi}}\right)\right),$$

which satisfies $\mathrm{P}\left(\mu_i\in I_\alpha^C(\widehat{\boldsymbol{\phi}})\right)=1-\alpha+o\left(m^{-1}\right)$.

Example 7.3 (*Estimating poverty indicators in Spain*) Sugasawa and Kubokawa (2019) applied the transformed nested error regression model to the estimation of poverty indicators in Spanish provinces, using the synthetic income data available in the sae package in R. The dataset consists of samples in $m=52$ areas, and the sample sizes (the number of observed units) range from 20 to 1420, and the total number of sample units is 17199. The welfare variable for the individuals is the equivalized annual net income denoted by E_{ij}, noting that the small portions of E_{ij} take negative values. As auxiliary variables, there are indicators of the four groupings of ages (16–24, 25–49, 50–64, and ≥65), the indicator of having Spanish nationality, the indicators of education levels (primary education and post-secondary education), and the indicators of two employment categories (employed and unemployed). Sugasawa and Kubokawa (2019) adopted four transformations, shifted DP (SDP) transformation, SDP transformation with known shift (SDP-s), sinh-arcsinh transformation, and shifted log transformation, where fixed shifted parameter is set to $c^*=|\min(E_{ij})|+1$. The estimated parameters in SDP are $\widehat{\lambda}=0.090$ (standard error is 1.99×10^{-3}) and $\widehat{c} = 4319$ (standard error is 170.69), which shows that the log-transformation is not suitable for this dataset. It is also shown that the SDP transformation attains the minimum value of the information criteria among the four transformations. See Sugasawa and Kubokawa (2019) for other results on point and interval estimates of area-wise poverty indicators.

7.5 Models with Skewed Distributions

The normality assumption for both random effects and error terms in the basic small area models is not plausible for response variables having skewed distributions. In the Fay–Herriot model (4.1), the normality assumption of the error term ε_i is based on central limit theorems since y_i is typically a summary statistic. However, when the sample size for computing y_i is small, the normality assumption would be violated. To address this issue, Ferraz and Mourab (2012) adopted a skew-normal distribution for the error term. Here, we focus on the use of skew-normal distribution in the nested error regression model.

The skew-normal distribution $\mathrm{SN}(\mu, \sigma^2, \lambda)$ is a distribution with density

$$f(x; \mu, \sigma^2, \lambda) = \frac{2}{\sigma}\phi\left(\frac{x-\mu}{\sigma}\right)\Phi\left(\lambda\frac{x-\mu}{\sigma}\right),$$

where $\phi(\cdot)$ and $\Phi(\cdot)$ being the density and distribution functions of the standard normal distribution. Here μ and σ are location and scale parameters, respectively, and λ controls the asymmetry of the distribution and varies in $(-\infty, \infty)$. An attractive feature of the above skew-normal distribution is that it includes the normal distribution $\mathrm{N}(\mu, \sigma^2)$ can be obtained as a special case by setting $\lambda = 0$. On the other hand, the distribution converges to the half-normal distribution under $\lambda \to \infty$. Furthermore, the skew-normal distribution admits a stochastic representation

$$Y \sim \mathrm{SN}(\mu, \sigma^2, \lambda) \ \Leftrightarrow \ Y = \mu + \sigma Z, \quad Z = \delta X_0 + \sqrt{1-\delta^2}X_1,$$

where $\delta = \lambda/\sqrt{1+\lambda^2}$, $X_0 \sim \mathrm{N}(0, 1)$ and $X_1 \sim N_+(0, 1)$ (truncated normal distribution on the positive real line). Note that, for $Y \sim \mathrm{SN}(\mu, \sigma^2, \lambda)$, it follows that $E(Y) = \mu + \delta\sigma\sqrt{2/\pi}$. Hence, the distribution $\mathrm{SN}(-\delta\sigma\sqrt{2/\pi}, \sigma^2, \lambda)$ has zero mean and can be used for skewed random errors.

Diallo and Rao (2018) extended the nested error regression models to have skewed random effects and error terms. The proposed model is

$$v_i \sim \mathrm{SN}(-\delta_v\tau\sqrt{2/\pi}, \tau^2, \lambda_v), \qquad \varepsilon_{ij} \sim \mathrm{SN}(-\delta_\varepsilon\sigma\sqrt{2/\pi}, \sigma^2, \lambda_\varepsilon)$$

where $\delta_v = \lambda_v/\sqrt{1+\lambda_v^2}$ and $\delta_\varepsilon = \lambda_\varepsilon/\sqrt{1+\lambda_\varepsilon^2}$. Note that the above settings ensure that $\mathrm{E}[v_i] = 0$ and $\mathrm{E}[\varepsilon_{ij}] = 0$. Diallo and Rao (2018) showed that the marginal distribution of $\boldsymbol{y}_i$ belongs to a family of the closed skew-normal distribution, which can enable us to conduct parameter estimation via the maximum likelihood method and derive the conditional prediction of non-sampled units.

Here, we focus on a sub-model of the skew-normal nested error regression model, defined by setting $\lambda_v = 0$, as explored in Tsujino and Kubokawa (2019). The model is described as

$$\boldsymbol{y}_i = \boldsymbol{X}_i\boldsymbol{\beta} + \mathbf{1}_{n_i}v_i + \boldsymbol{\epsilon}_i, \quad i = 1, \ldots, m,$$

where $v_i \sim \mathrm{N}(0, \tau^2)$ and

$$\boldsymbol{\epsilon}_i = \frac{\sigma}{\sqrt{1+\lambda^2}}\boldsymbol{u}_{0i} + \frac{\sigma\lambda}{\sqrt{1+\lambda^2}}\boldsymbol{u}_{1i}.$$

Here, $\boldsymbol{u}_{0i} = (u_{0i1}, \ldots, u_{0in_i})$ and $\boldsymbol{u}_{1i} = (u_{1i1}, \ldots, u_{1in_i})$, where u_{0ij} and u_{1ij} are independent and $u_{0ij} \sim \mathrm{N}(0, 1)$ and $u_{1ij} \sim N_+(0, 1)$. We first find the Bayes estimator of v_i. The conditional distribution of y_{ij} given v_i and u_{1ij} is given by

$$f\left(y_{ij} \mid v_i, u_{1ij}\right) = \phi\left(y_{ij}; x_{ij}^\top\beta + v_i + \frac{\sigma\lambda}{\sqrt{1+\lambda^2}}u_{1ij}, \frac{\sigma^2}{1+\lambda^2}\right).$$

Then, the conditional distribution of $(v_i, \boldsymbol{u}_{1i})$ given $\boldsymbol{y}_i$ is

$$f\left(v_i, \boldsymbol{u}_{1i} \mid \boldsymbol{y}_i\right)$$
$$\propto \left\{\prod_{j=1}^{n_i}\phi\left(y_{ij}; x_{ij}^\top\beta + v_i + \frac{\sigma\lambda}{\sqrt{1+\lambda^2}}u_{1ij}, \frac{\sigma^2}{1+\lambda^2}\right)\phi_+(u_{ij}; 0, 1)\right\}\phi(v_i; 0, \tau^2),$$

where $\phi_+(u_{ij}; a, b)$ denotes the density of the truncated normal distribution on the positive real line with mean and variance parameters, a and b, respectively. Define

$$\mu_{v_i} = \frac{n_i\tau^2\left(1+\lambda^2\right)}{\sigma^2 + n_i\tau^2\left(1+\lambda^2\right)}\left(\bar{y}_i - \bar{\boldsymbol{x}}_i^\top\boldsymbol{\beta} - \frac{\sigma\lambda}{\sqrt{1+\lambda^2}}n_i^{-1}\sum_{j=1}^{n_i}u_{1ij}\right),$$

$$s_{v_i}^2 = \frac{\sigma^2\tau^2}{\sigma^2 + n_i\tau^2\left(1+\lambda^2\right)}.$$

Also, let $\boldsymbol{R}_i = (1-\rho_i)\boldsymbol{I}_{n_i} + \rho_i\mathbf{1}_{n_i}\mathbf{1}_{n_i}^\top$ for the $n \times n$ identity matrix $\boldsymbol{I}_n$ and

$$\rho_i = \frac{\tau^2\lambda^2/\left(\sigma^2 + n_i\tau^2\right)}{1 + \tau^2\lambda^2/\left(\sigma^2 + n_i\tau^2\right)}.$$

Denote the (j, k) element of $\boldsymbol{R}_i$ by $r_{i,jk}$, namely, $r_{i,jk} = 1$ for $j = k$ and $r_{i,jk} = \rho_i$ for $j \neq k$. Then, the conditional density can be rewritten as

$$f\left(v_i, \boldsymbol{u}_{1i} \mid \boldsymbol{y}_i\right) \propto \phi\left(v_i; \mu_{v_i}, s_{v_i}^2\right) \phi_{n_i}\left(\boldsymbol{u}_{1i}; \boldsymbol{\xi}_i, s_{u_i}^2 \boldsymbol{R}_i\right) \prod_{i=1}^{n_i} I\left(u_{1ij} > 0\right)$$

where $\boldsymbol{\xi}_i = \left(\xi_{i1}, \dots, \xi_{in_i}\right)^\top$ for

$$\xi_{ij} = \frac{\lambda}{\sigma\sqrt{1+\lambda^2}}\left\{y_{ij} - \boldsymbol{x}_{ij}^\top\boldsymbol{\beta} - \frac{n_i\tau^2}{\sigma^2 + n_i\tau^2}\left(\bar{y}_i - \bar{\boldsymbol{x}}_i^\top\boldsymbol{\beta}\right)\right\}$$

and

$$s_{u_i}^2 = \frac{1}{1+\lambda^2}\left(1 + \frac{\tau^2\lambda^2}{\sigma^2 + n_i\tau^2}\right).$$

Let $\boldsymbol{w}_i = \left(w_{i1}, \dots, w_{in_i}\right)^\top = \left(\boldsymbol{u}_{1i} - \boldsymbol{\xi}_i\right)/s_{u_i}$. Then, it holds that $\boldsymbol{w}_i|\boldsymbol{y}_i \sim \mathrm{TN}_{\mathbb{R}_+^{n_i}}$ $(\boldsymbol{\xi}_i/a_{u_i}, \boldsymbol{R}_i)$, where $\mathrm{TN}_{\mathbb{R}_+^{n_i}}(\boldsymbol{A}, \boldsymbol{B})$ is the multivariate truncated normal distribution on $(0, \infty)^{n_i}$ with mean and covariance parameters, $\boldsymbol{A}$ and $\boldsymbol{B}$, respectively. Then, we have

$$\begin{aligned}\widetilde{v}_i(\boldsymbol{\phi}) &\equiv \mathrm{E}\left[v_i \mid \boldsymbol{y}_i\right] = \mathrm{E}\left[\mathrm{E}\left[v_i \mid \boldsymbol{y}_i, \boldsymbol{u}_{1i}\right] \mid \boldsymbol{y}_i\right] \\ &= \frac{n_i\tau^2}{\sigma^2 + n_i\tau^2}\left(\bar{y}_i - \bar{\boldsymbol{x}}_i^\top\boldsymbol{\beta}\right) - \frac{\sigma\lambda}{\sqrt{1+\lambda^2}}\frac{n_i\tau^2\left(1+\lambda^2\right)}{\sigma^2 + n_i\tau^2\left(1+\lambda^2\right)} s_{u_i} n_i^{-1}\sum_{j=1}^{n_i}\mathrm{E}\left[w_{ij} \mid \boldsymbol{y}_i\right],\end{aligned}$$

where $\boldsymbol{\phi} = (\boldsymbol{\beta}, \tau^2, \sigma^2, \lambda_\varepsilon)$ is a collection of the unknown parameters. To obtain the expression of $\widetilde{v}_i(\boldsymbol{\phi})$, we need to compute the conditional expectation of w_{ij} given $\boldsymbol{y}_i$, which includes n_i-dimensional integral. However, a simplified expression for the expectation is provided in Tsujino and Kubokawa (2019).

Regarding the estimation of $\boldsymbol{\phi}$, we first consider the estimation $\boldsymbol{\beta}$ with the other parameters fixed. We note that

$$\mathrm{E}[y_{ij}] = \boldsymbol{x}_{ij}^\top\boldsymbol{\beta} + \mu_\varepsilon, \qquad \mu_\varepsilon = \sigma\sqrt{\frac{2}{\pi}}\frac{\lambda}{\sqrt{1+\lambda^2}}$$

and

$$\boldsymbol{V}_i \equiv \mathrm{Var}(\boldsymbol{y}_i) = \sigma^2\left(1 - \frac{2}{\pi}\frac{\lambda^2}{1+\lambda^2}\right)\boldsymbol{I}_{n_i} + \tau^2\boldsymbol{J}_{n_i}.$$

Let $\boldsymbol{\beta}_\varepsilon = (\beta_0 + \mu_\varepsilon, \beta_1, \dots, \boldsymbol{\beta}_p)$, where β_0 is an intercept parameter. Then, the generalized least squares estimator of $\boldsymbol{\beta}_\varepsilon$ is

$$\widetilde{\boldsymbol{\beta}}_\varepsilon = \left(\sum_{i=1}^m \boldsymbol{X}_i^\top\boldsymbol{V}_i^{-1}\boldsymbol{X}_i\right)^{-1}\sum_{i=1}^m \boldsymbol{X}_i^\top\boldsymbol{V}_i^{-1}\boldsymbol{y}_i,$$

so that the estimator $\widetilde{\boldsymbol{\beta}}$ of $\boldsymbol{\beta}$ is obtained as $\widetilde{\boldsymbol{\beta}} = \widetilde{\boldsymbol{\beta}}_{\varepsilon} - (\mu_{\varepsilon}, \mathbf{0}_p^{\top})$. For the estimation of σ^2, τ^2 and λ, Tsujino and Kubokawa (2019) proposed a moment-based estimator, but the maximum likelihood estimator discussed in Diallo and Rao (2018) can also be adopted.

References

Azzalini A (2005) The skew-normal distribution and related multivariate families (with discussion). Scand J Stat 32:159–188

Ferraz V, Mourab FAS (2012) Small area estimation using skew normal models. Comput Stat Data Anal 56:2864–2874

Ghosh M, Maiti T (2004) Small-area estimation based on natural exponential family quadratic variance function models and survey weights. Biometrika 91:95–112

Ghosh M, Maiti T (2008) Empirical Bayes confidence intervals for means of natural exponential family-quadratic variance function distributions with application to small area estimation. Scand J Stat 35:484–495

Ghosh M, Natarajan K, Stroud TWF, Carlin BP (1998) Generalized linear models for small area estimation. J Am Stat Assoc 93:273–282

Hirose MY, Ghosh M, Ghosh T (2023) Arc-sin transformation for binomial sample proportions in small area estimation. Stat Sin (to appear)

Jiang J, Lahiri P (2001) Empirical best prediction for small area inference with binary data. Ann Inst Stat Math 53:217–243

Jiang J, Nguyen T (2007) Linear and generalized linear mixed models and their applications. Springer, New York

Molina I, Martin N (2018) Empirical best prediction under a nested error model with log transformation. Ann Stat 46:1961–1993

Slud E, Maiti T (2006) Mean-squared error estimation in transformed Fay-Herriot models. J Roy Stat Soc B 68:239–257

Sugasawa S, Kubokawa T (2015) Parametric transformed Fay-Herriot model for small area estimation. J Multivar Anal 139:17–33

Sugasawa S, Kubokawa T, Rao JNK (2018) Small area estimation via unmatched sampling and linking models. Test 27:407–427

Sugasawa S, Kubokawa T, Rao JNK (2019) Hierarchical Bayes small area estimation with an unknown link function. Scand J Stat 46:885–897

Sugasawa S, Kubokawa T (2017) Transforming response values in small area prediction. Comput Stat Data Anal 114:47–60

Sugasawa S, Kubokawa T (2019) Adaptively transformed mixed model prediction of general finite population parameters. Scand J Stat 46:1025–1046

Tsujino T, Kubokawa T (2019) Empirical Bayes methods in nested error regression models with skew-normal errors. Jpn J Stat Data Sci 2:375–403

Yang ZL (2006) A modified family of power transformations. Econ Lett 92:14–19

You Y, Rao JNK (2002) Small area estimation using unmatched sampling and linking models. Can J Stat 30:3–15

Chapter 8
Extensions of Basic Small Area Models

The flexibility of the two basic small area models described in Chap. 4 can be limited for practical applications. To overcome the difficulty, we here introduce some extensions of the basic small area models. Specifically, we focus on flexible modeling of random effects, measurement errors in covariates, nonparametric and semiparametric modeling, and modeling heteroscedastic variance.

8.1 Flexible Modeling of Random Effects

As discussed in Chap. 4, the random effects play crucial roles to express area-wise variability in small area estimation. For modeling random effects, normal distributions are typically used due to their computational convenience. However, it may not be the case in practice, and the misspecification of the random effects distribution may lead to inefficient small area estimation.

8.1.1 Uncertainty of the Presence of Random Effects

The first issue is regarding the model selection regarding the inclusion of random effects. The importance of the preliminary testing for the presence of random effects is addressed in Datta and Mandal (2015) and Molina et al. (2015). These papers demonstrated that eliminating the random effects can improve the accuracy of the small area estimators when the random effects are not necessary. Here, we consider a probabilistic framework for the model selection, that is, we incorporate uncertainty of the existence of random effects into the mixed-effects model. Datta and Mandal

S. Sugasawa and T. Kubokawa, *Mixed-Effects Models and Small Area Estimation*, JSS Research Series in Statistics, https://doi.org/10.1007/978-981-19-9486-9_8

(2015) proposed the Fay–Herriot model (4.1) with random effect v_i following the two-component mixture distribution, described as

$$v_i \sim p\mathrm{N}(0, A) + (1 - p)\delta_0, \quad i = 1, \ldots, m, \tag{8.1}$$

where δ_0 is the one-point distribution on the origin. Here, p is an unknown parameter representing the prior probability of existence of random effects in the ith area. The second mixture component is the one-point distribution on the origin, representing the situation that the random effect in the ith area is not necessary. Note that the model (8.1) is quite similar to the spike-and-slab prior (e.g., Scott and Berger 2006) for variable selection. The above model (8.1) can be expressed in the following hierarchy:

$$v_i|(s_i = 1) \sim \mathrm{N}(0, A), \quad v_i|(s_i = 0) \sim \delta_0, \quad \mathrm{P}(s_i = 1) = p,$$

where s_i is a latent random variable indicating whether random area effect should be needed ($s_i = 1$) or not ($s_i = 0$). Unlike the preliminary test method, the probabilistic formulation (8.1) allows for the coexistence of areas with and without random effects so that Datta and Mandal (2015) called 'uncertain random effects' for the random effects structure of (8.1). Under the model (8.1), the marginal distribution of the observed value y_i is given by

$$y_i \sim p\mathrm{N}(\boldsymbol{x}_i^\top \boldsymbol{\beta}, A + D_i) + (1 - p)\mathrm{N}(\boldsymbol{x}_i^\top \boldsymbol{\beta}, D_i),$$

which is a mixture distribution of two marginal distributions with and without random effects.

Using the hierarchical expression of the model, the conditional expectation of $\theta_i \equiv \boldsymbol{x}_i^\top \boldsymbol{\beta} + v_i$ is obtained as

$$\begin{aligned}\mathrm{E}[\theta_i|y_i] &= \mathrm{E}[\theta_i|y_i, s_i = 1]\mathrm{P}(s_i = 1|y_i) + \mathrm{E}[\theta_i|y_i, s_i = 0]\mathrm{P}(s_i = 0|y_i)\\ &= \boldsymbol{x}_i^\top \boldsymbol{\beta} + \frac{A}{A + D_i}(y_i - \boldsymbol{x}_i^\top \boldsymbol{\beta})\widetilde{p}_i(y_i; \boldsymbol{\beta}, A),\end{aligned}$$

where

$$\widetilde{p}_i(y_i; \boldsymbol{\beta}, A) = \frac{p}{p + (1 - p)\sqrt{\frac{A+D_i}{D_i}}\exp\left\{-\frac{1}{2}\frac{A(y_i - \boldsymbol{x}_i^\top \boldsymbol{\beta})^2}{D_i(A+D_i)}\right\}} \tag{8.2}$$

is the conditional probability being $s_i = 1$. It is observed that $\mathrm{E}[\theta_i|y_i]$ is a weighted average of two conditional expectations given $s_i = 1$ and $s_i = 0$ so that the uncertainty of the presence of random effects is reflected in the estimator $\mathrm{E}[\theta_i|y_i]$.

For estimating the unknown parameters, Datta and Mandal (2015) proposed a hierarchical Bayesian approach by assigning prior distributions on the parameters. Specifically, they employed the uniform prior for $\boldsymbol{\beta}$, a proper inverse gamma prior

IG(a, b) for A and a beta prior Be(c, d) for p. Then, we get the joint posterior distribution given by

$$\pi\left(\boldsymbol{v}, \boldsymbol{s}, \boldsymbol{\beta}, A, p \mid \boldsymbol{y}\right)$$

$$\propto \exp\left\{-\frac{1}{2}\sum_{i=1}^{m}\frac{\left(y_i - \boldsymbol{x}_i^\top\boldsymbol{\beta} - v_i\right)^2}{D_i} - \frac{1}{2A}\sum_{i=1}^{m}s_i v_i^2\right\}\prod_{i=1}^{m}A^{-s_i/2}\left\{I\left(v_i = 0\right)\right\}^{1-s_i}$$

$$\times\, p^{c+\sum_{i=1}^{m}s_i-1}(1-p)^{d+m-\sum_{i=1}^{m}s_i-1}A^{-b-1}\exp\left(-\frac{a}{A}\right).$$

The propriety of the above posterior density is proved in Datta and Mandal (2015) under reasonable conditions. The posterior distribution can be approximated by the Markov Chain Monte Carlo algorithm. In particular, since the full conditional distribution of each parameter is a familiar form, the posterior samples can be generated by a simple Gibbs sampler. The list of full conditional distributions is given as follows:

- $\boldsymbol{\beta}|\boldsymbol{v}, A, \boldsymbol{y} \sim \mathrm{N}(\{\sum_{i=1}^{m}(A+D_i)^{-1}\boldsymbol{x}_i\boldsymbol{x}_i^\top\}^{-1}\sum_{i=1}^{m}(A+D_i)^{-1}\boldsymbol{x}_i(y_i - v_i),$ $\{\sum_{i=1}^{m}(A+D_i)^{-1}\boldsymbol{x}_i\boldsymbol{x}_i^\top\}^{-1}),$
- $v_i|s_i, \boldsymbol{\beta}, A, p, y_i \sim \delta_0$ if $s_i = 0$, and $v_i|s_i, \beta, \pi, A, y_i \sim \delta_0 \sim \mathrm{N}(A(y_i - \boldsymbol{x}_i^\top\beta)/(A+D_i), AD_i/(A+D_i))$ if $s_i = 1$, for $i = 1, \ldots, m$,
- $s_i|y_i, \boldsymbol{\beta}, A \sim \mathrm{Ber}(\widetilde{p}_i(y_i; \boldsymbol{\beta}, A))$, where $\widetilde{p}_i(y_i; \boldsymbol{\beta}, A)$ is given in (8.2), for $i = 1, \ldots, m$,
- $p|\boldsymbol{s} \sim \mathrm{Be}(c + \sum_{i=1}^{m}s_i, d + m - \sum_{i=1}^{m}s_i)$,
- $A|\boldsymbol{v} \sim \mathrm{IG}(a + m/2, b + \sum_{i=1}^{m}v_i^2/2)$.

Starting with some initial values, the Gibbs sampler iteratively sample from the above full conditional distributions. Based on the posterior samples of θ_i, the hierarchical Bayes estimator is obtained as the posterior mean.

As an extension of the uncertain random effects under the Fay–Herriot model, Sugasawa et al. (2017) proposed 'uncertain empirical Bayes' for the area-level model based on the natural exponential family described in Sect. 7.2. Remember that the conditional distribution of y_i given θ_i is

$$y_i \mid \theta_i \sim f\left(y_i \mid \theta_i\right) = \exp\left\{n_i\left(\theta_i y_i - \psi\left(\theta_i\right)\right) + c\left(y_i, n_i\right)\right\},$$

where n_i is a known scalar value in the ith area. Note that the small area parameter of interest is the conditional mean $\mu_i \equiv \mathrm{E}[y_i|\theta_i] = \psi'(\theta_i)$. To express the uncertainty of the presence of random effects, the following two-component mixture distribution is introduced:

$$\theta_i|(s_i = 1) \sim \pi\left(\theta_i\right) = \exp\left\{\nu\left(m_i\theta_i - \psi\left(\theta_i\right)\right) + C\left(\nu, m_i\right)\right\},$$
$$\theta_i|(s_i = 0) = (\psi')^{-1}(m_i),$$

where $\mathrm{P}(s_i = 1) = 1 - \mathrm{P}(s_i = 0) = p$, s_i is an indicator of the presence of random area effect, m_i is the mean of the conjugate prior and $(\psi')^{-1}$ denotes the inverse function of ψ' given in Sect. 7.2. Since $\mu_i|(s_i = 0) = m_i$, the prior distribution of the small area mean μ_i reduces to the one-point distribution on the regression part m_i when there is no random effect in the ith area. The joint density (or mass) function of (y_i, θ_i, s_i) is

$$g(y_i, \theta_i, s_i = 1) = f(y_i \mid \theta_i)\pi(\theta_i),$$
$$g(y_i, \theta_i, s_i = 0) = \delta_{\theta_i}\left(\left(\psi'\right)^{-1}(m_i)\right) f(y_i \mid \theta_i),$$

where $\delta_{\theta_i}(a)$ denotes the point mass at $\theta_i = a$. Then the marginal distribution of y_i is a mixture of two distributions:

$$f(y_i; \boldsymbol{\phi}) = p f_1(y_i; \boldsymbol{\phi}) + (1-p) f_2(y_i; \boldsymbol{\phi}),$$

where $\boldsymbol{\phi}$ is a collection of unknown parameters and

$$f_1(y_i; \boldsymbol{\phi}) = \int f(y_i \mid \theta_i)\pi(\theta_i)\, d\theta_i, \quad f_2(y_i; \boldsymbol{\phi}) = f\left(y_i \mid \theta_i = \left(\psi'\right)^{-1}(m_i)\right).$$

Since $\pi(\theta_i)$ is the conjugate prior of θ_i, the marginal distribution $f_1(y_i; \boldsymbol{\phi})$ can be obtained in a closed form. The conditional distribution of $s_i = 1$ given y_i can be obtained as

$$P(s_i = 1 \mid y_i; \boldsymbol{\phi}) \equiv r_i(y_i; \boldsymbol{\phi}) = \frac{p}{p + (1-p) f_2(y_i; \boldsymbol{\phi}) / f_1(y_i; \boldsymbol{\phi})}.$$

Then, the conditional expectation of μ_i given y_i is

$$\widetilde{\mu}_i(y_i; \boldsymbol{\phi}) = m_i + \frac{n_i}{\nu + n_i}(y_i - m_i) r_i(y_i; \boldsymbol{\phi}).$$

It should be noted that the estimator $\widetilde{\mu}_i(y_i; \boldsymbol{\phi})$ reduces to the estimator under the model with uncertain random effect proposed by Datta and Mandal (2015), when the conditional distribution of y_i given θ_i is $\mathrm{N}(\theta_i, D_i)$. Therefore, the mixture prior for θ_i is a reasonable extension of the uncertain random effect to handle general response variables.

To estimate the unknown parameter $\boldsymbol{\phi}$, Sugasawa et al. (2017) proposed the Monte Carlo EM algorithm to maximize the marginal likelihood. The detailed steps of the algorithm are presented as follows:

1. Set the initial value $\boldsymbol{\phi}^{(0)}$ and $r = 0$.
2. Compute $a_i^{(r)} \equiv \mathrm{E}^{(r)}[\theta_i | s_i = 1]$ and $b_i^{(r)} \equiv \mathrm{E}^{(r)}[\psi(\theta_i) | s_i = 1]$, where $\mathrm{E}^{(r)}$ denotes the expectation with the current parameter value $\boldsymbol{\phi}^{(r)}$.
3. Update the parameter values as follows:

$$\left(\boldsymbol{\beta}^{(r+1)}, \nu^{(r+1)}\right) = \text{argmax}_{\boldsymbol{\beta},\nu} \sum_{i=1}^{m} r_i\left(y_i, \boldsymbol{\phi}^{(r)}\right)\left\{\nu m_i a_i^{(r)} - \nu b_i^{(r)} + C\left(\nu, m_i\right)\right\}$$

$$p^{(r+1)} = \frac{1}{m}\sum_{i=1}^{m} r_i\left(y_i, \boldsymbol{\phi}^{(r)}\right)$$

4. If the difference between $\boldsymbol{\phi}^{(r)}$ and $\boldsymbol{\phi}^{(r+1)}$ is sufficiently small, the output is $\boldsymbol{\phi}^{(r+1)}$. Otherwise, set $r = r + 1$ and go back to Step 2.

Substituting $\widehat{\boldsymbol{\phi}}$ obtained from the above algorithm, we finally obtain the empirical uncertain Bayes (EUB) estimator of μ_i as $\widetilde{\mu}_i(y_i; \widehat{\boldsymbol{\phi}})$. For measuring uncertainty of the EUB estimator, Sugasawa et al. (2017) derived the second-order unbiased estimator of conditional MSE.

A representative model included in the exponential family is the Poisson-gamma with uncertain random effects (UPG) model, described as

$$z_i \sim \text{Po}(n_i\lambda_i), \quad \lambda_i \sim p\text{Ga}(\nu m_i, \nu) + (1-p)\delta_{m_i}, \quad i = 1, \ldots, m, \tag{8.3}$$

where $z_i = n_i\lambda_i$, δ_{m_i} is the one-point distribution on $\lambda_i = m_i$, and $m_i = \exp(\boldsymbol{x}_i^\top \boldsymbol{\beta})$. This model is useful for modeling disease/mortality risk as used in the following example.

Example 8.1 (*Historical mortality data in Tokyo*) Sugasawa et al. (2017) applied the uncertain empirical Bayes method to historical mortality data in Tokyo. The mortality rate is a representative index in demographics and has been used in various fields. Especially, in economic history, one can discover new knowledge from a spatial distribution of mortality rate in small areas. However, the direct estimate of the mortality rate in small areas with extremely low population has high variability, which may lead to incorrect recognition of the spatial distribution. Therefore, it is desirable to use smoothed and stabilized estimates through empirical Bayes methods. The dataset is the mortality data in Tokyo in 1930 and consists of the observed mortalities z_i and the number of population N_i in the ith area in Tokyo. Such area-level data are available for $m = 1371$ small areas. Let n_i be the standardized mortality ratio obtained as $n_i = N_i \sum_{i=1}^m z_i / \sum_{i=1}^m N_i$. For this dataset, the UPG model (8.3) as well as the standard Poisson-gamma (PG) model are applied. The estimate of p is 0.56, indicating that random effects are not necessarily required around half of the areas. Furthermore, AIC of UPG is 8142 and that of PG is 8265, so that the UPG model fits to the dataset better than the PG model.

Finally, we note that there are other extensions of the uncertain random effect proposed by Datta and Mandal (2015). Sugasawa and Kubokawa (2017a) employed the formulation (8.1) in the nested error regression model and demonstrated the usefulness in the context of estimating parameters in a finite population framework. Furthermore, Chakraborty et al. (2016) proposed a two component mixture model, namely, $v_i \sim \pi\text{N}(0, A_1) + (1-\pi)\text{N}(0, A_2)$ with unknown A_1 and A_2.

8.1.2 Modeling Random Effects via Global–Local Shrinkage Priors

A new direction for modeling random effects is the use of the global–local shrinkage prior originally developed in the context of signal estimation (e.g., Carvalho et al. (2010)). The first work to cast the global–local shrinkage prior in the context of small area estimation is Tang et al. (2018). The authors proposed the extension of the Fay–Herriot model by using the following specification of the random effect:

$$v_i|\lambda_i^2, \tau^2 \sim \mathrm{N}(0, \lambda_i^2 A), \quad \lambda_i^2 \sim \pi(\lambda_i^2).$$

Given λ_i^2, the conditional expectation of θ_i is

$$\mathrm{E}[\theta_i|y_i, \lambda_i^2] = y_i - B_i(y_i - \boldsymbol{x}_i^\top \boldsymbol{\beta}), \quad B_i = \frac{D_i}{D_i + \lambda_i^2 A}$$

so that the shrinkage factor B_i depends on the area-specific parameter λ_i^2, which adds flexibility to the shrinkage estimation.

There are several choices for the distribution of λ_i^2. For example, the exponential distribution for λ_i^2 leads to the Laplace prior for v_i as the marginal distribution. Another representative example is the half-Cauchy distribution for λ_i (equivalent to $\pi(\lambda_i^2) \propto (\lambda_i^2)^{-1/2}(1 + \lambda_i^2)^{-1}$), leading to the horseshoe prior (Carvalho et al. 2010) for v_i. Other examples are listed in Table 1 in Tang et al. (2018).

An important property of the global–local shrinkage prior is that the shrinkage factor can change depending on the observed value y_i. To see the property, we consider the situation, $|y_i - \boldsymbol{x}_i^\top \boldsymbol{\beta}| \to \infty$, under which the regression part $\boldsymbol{x}_i^\top \boldsymbol{\beta}$ is not useful to improve the accuracy of the direct estimator y_i. In this case, it is not a good idea to shrinkage y_i toward $\boldsymbol{x}_i^\top \boldsymbol{\beta}$ as in the standard small area estimator. Tang et al. (2018) showed that $\mathrm{P}(B_i > \varepsilon|y_i, \boldsymbol{\beta}, A) \to 0$ if $|y_i - \boldsymbol{x}_i^\top \boldsymbol{\beta}| \to \infty$ and the distribution of λ_i^2 satisfies some conditions.

To estimate the model parameters as well as the small area parameters, Tang et al. (2018) proposed a hierarchical Bayes approach. Suppose that prior distributions for β and A are the uniform distribution and $\mathrm{IG}(c, d)$, respectively. The joint posterior density under the global–local shrinkage prior is given by

$$\pi\left(\boldsymbol{\beta}, A, \boldsymbol{v}, \boldsymbol{\lambda} \mid \boldsymbol{y}\right) \propto \exp\left\{-\frac{1}{2}\sum_{i=1}^{m} \frac{\left(y_i - \boldsymbol{x}_i^\top \boldsymbol{\beta} - v_i\right)^2}{D_i} - \frac{1}{2}\sum_{i=1}^{m} \frac{v_i^2}{\lambda_i^2 A} - \frac{d}{A}\right\}$$
$$\times A^{-m/2-c-1} \prod_{i=1}^{m} \pi(\lambda_i^2)(\lambda_i^2)^{-\frac{1}{2}}.$$

We can use a Gibbs sampler to generate posterior samples from the above density, where the full conditional distributions are obtained as follows:

- $v_i|\beta, A, \lambda_i^2, y_i \sim \mathrm{N}((1 - B_i)(y_i - \boldsymbol{x}_i^\top \boldsymbol{\beta}), (1 - B_i)D_i)$ for $i = 1, \ldots, m$,
- $\boldsymbol{\beta}|\boldsymbol{v}, \boldsymbol{y} \sim \mathrm{N}((\sum_{i=1}^m D_i^{-1}\boldsymbol{x}_i\boldsymbol{x}_i^\top)^{-1} \sum_{i=1}^m D_i^{-1}\boldsymbol{x}_i(y_i - v_i), (\sum_{i=1}^m D_i^{-1}\boldsymbol{x}_i\boldsymbol{x}_i^\top)^{-1})$,
- $A|\boldsymbol{v}, \boldsymbol{\lambda} \sim \mathrm{IG}(c + m/2, d + \sum_{i=1}^m v_i^2/2\lambda_i^2)$,
- $\pi(\lambda_i^2|A, v_i) \propto \pi(\lambda_i^2)(\lambda_i^2)^{-1/2}\exp(-v_i^2/2\lambda_i^2 A)$ for $i = 1, \ldots, m$.

Hence, samples of $\boldsymbol{\beta}$, A and v_i can be directly drawn from familiar distributions. To draw samples of λ_i^2, one may need to use a Metropolis–Hastings step in general. However, in some cases, the full conditional becomes a familiar distribution. For example, when one uses the exponential prior for λ_i^2, the full conditional is $\pi(\lambda_i^2|A, v_i) \propto (\lambda_i^2)^{-1/2}\exp(-v_i^2/2\lambda_i^2 A - \lambda_i^2)$, which is the generalized inverse Gaussian (GIG) distribution. Furthermore, suppose that one uses the half-Cauchy prior for λ_i. Introducing a latent parameter ξ_i with $\lambda_i^2|\xi_i \sim \mathrm{Ga}(1/2, \xi_i)$ and $\xi_i \sim \mathrm{Ga}(1/2, 1)$, the full conditional of λ_i^2 is $\pi(\lambda_i^2|v_i, \xi_i) \propto (\lambda_i^2)^{-1/2}\exp(-v_i^2/2\lambda_i^2 A - \xi_i\lambda_i^2)$ (GIG distribution) and $\xi_i|\lambda_i^2 \sim \mathrm{Ga}(1, 1 + \lambda_i^2)$.

Example 8.2 (*State-level child poverty ratio*) Tang et al. (2018) applied the Fay–Herriot model with the horseshoe and Laplace priors to the state-level direct estimates of the poverty ratio for the age group 5–17 that were obtained from the 1999 Current Population Survey. The response variable is the direct estimator of the poverty ratio and three covariates are available. The dataset is previously analyzed in Datta and Mandal (2015) using uncertain random effects and the presence of random effects was necessary only for Massachusetts. The posterior means of random effects v_i under the horseshoe prior were closer to zero than those from the standard Fay–Herriot model (i.e., normal prior for v_i), due to the adaptive and strong shrinkage properties of the horseshoe prior. To compare the accuracy of the estimates, the state-level poverty ratios obtained from the 2000 census are used as the true values and deviation measures such as absolute deviation and absolute relative bias are calculated for estimates obtained based on observed data in the 1999 survey. The results showed that the Laplace prior provides the smallest deviation measures among the candidates including the uncertain random effects method. See Tang et al. (2018) for more details of the results.

There are some attempts to extend the global–local shrinkage priors to non-normal settings. For count data, Datta and Dunson (2016) and Hamura et al. (2022) considered the following hierarchical model:

$$y_i|\lambda_i \sim \mathrm{Po}(\eta_i\lambda_i), \quad \lambda_i|u_i \sim \mathrm{Ga}(\alpha, \beta/u_i), \quad u_i \sim \pi(u_i), \quad i = 1, \ldots, m \tag{8.4}$$

where $\eta_i = \exp(\boldsymbol{x}_i^\top \gamma)$, γ is a regression coefficient, α and β are unknown parameters, and u_i is a local parameter to control area-wise shrinkage. Under this model, the conditional expectation of λ_i is given by

$$\mathrm{E}[\lambda_i|y_i] = \mathrm{E}\left[\frac{u_i}{\beta + \eta_i u_i}(\alpha + y_i)\Big|y_i\right] = \frac{y_i}{\eta_i} - \mathrm{E}\left[\frac{\beta}{\beta + \eta_i u_i}\left(\frac{y_i}{\eta_i} - \frac{\alpha u_i}{\beta}\right)\Big|y_i\right], \tag{8.5}$$

so that u_i controls the amount of shrinkage of y_i/η_i toward the prior mean $\alpha u_i/\beta$. For the prior for u_i, Hamura et al. (2022) proposed the extremely heavy-tailed (EH) prior, defined by the following density:

$$\pi_{\rm EH}(u_i; \gamma) = \frac{\gamma}{1+u_i} \frac{1}{\{1+\log(1+u_i)\}^{1+\gamma}},$$

for $\gamma > 0$. A notable property of the EH prior is that it holds tail-robustness, namely, $|y_i - \mathrm{E}[\lambda_i|y_i]| \to 0$ as $y_i \to \infty$. This means that the posterior mean under the EH prior holds a shrinkage property similar to one under Gaussian response as given in Tang et al. (2018). Furthermore, the EH prior admits an integral expression which leads to a tractable posterior computation algorithm.

To fit the hierarchical Poisson model (8.4), Hamura et al. (2022) adopted a Bayesian approach by assigning prior, $\alpha \sim \mathrm{Ga}(a_\alpha, b_\alpha, \beta \sim \mathrm{Ga}(a_\beta, b_\beta)$, where $a_\alpha, b_\alpha, a_\beta$ and b_β are fixed hyperparameters. The sampling steps for each parameter and latent variable are given as follows:

- The full conditional of λ_i is Ga $(y_i + \alpha, \eta_i + \beta/u_i)$ and $\lambda_1, \ldots, \lambda_m$ are mutually independent.
- The full conditional of β is Ga $\left(m\alpha + a_\beta, \sum_{i=1}^m \lambda_i/u_i + b_\beta\right)$.
- The sampling of dispersion parameter α can be done in multiple steps. The conditional posterior density of α obtained by marginalizing λ_i out is proportional to

$$\psi_\alpha(\alpha) \prod_{i=1}^m \frac{\Gamma(y_i+\alpha)}{\Gamma(\alpha)} \left(\frac{\beta}{\beta+\eta_i u_i}\right)^\alpha$$

where $\psi_\alpha(\alpha)$ is the prior density of α. The integer-valued variable ν_i is considered a latent parameter that augments the model and allows Gibbs sampler. Thus, we need to sample from the full conditionals of α and ν_i via the following two steps:

1. If $y_i = 0$, then $\nu_i = 0$ with probability one. Otherwise, ν_i is expressed as the distributional equation $\nu_i = \sum_{j=1}^{y_i} d_j$, where d_j $(j = 1, \ldots, y_i)$ are independent random variables distributed as $\mathrm{Ber}(\alpha/(j-1+\alpha))$.
2. Generate α from Ga $\left(\sum_{i=1}^m \nu_i + a_\alpha, \sum_{i=1}^m \log(1+\eta_i u_i/\beta) + b_\alpha\right)$.

- The sampling from u_i is done with sampling from auxiliary latent variables. The details steps are given as follows:

1. Generate w_i from Ga $(1+\gamma, 1+\log(1+u_i))$.
2. Generate v_i from Ga $(1+w_i, 1+u_i)$.
3. Generate u_i from GIG $(1-\alpha, 2v_i, 2\beta\lambda_i)$, where $\mathrm{GIG}(a, b, p)$ is the generalized inverse Gaussian distribution with density $\pi(x; a, b, p) \propto x^{p-1}\exp\{-(ax+bx)/2\}$ for $x > 0$.

- Under gamma prior $\gamma \sim$ Ga (a_γ, b_γ), the full conditional of γ (with (v_i, w_i) marginalized out) is Ga $\left(a_\gamma + m, b_\gamma + \sum_{i=1}^m \log\{1+\log(1+u_i)\}\right)$.

Example 8.3 (*Crime risk estimate*) Hamura et al. (2022) applied the Poisson models with global–local shrinkage priors to crime count in $m = 2855$ local towns in the Tokyo metropolitan area in 2015. As auxiliary information in each town, area (km^2) and five covariates are available. Let y_i $(i = 1, \dots, m)$ be the observed crime count, a_i be the area, and x_i be the vector of the standardized auxiliary information. For the dataset, the following Poisson model is considered:

$$y_i \mid \lambda_i \sim \mathrm{Po}\left(\lambda_i \eta_i\right), \quad \eta_i = \exp\left(\log a_i + x_i^\top \delta\right), \quad \lambda_i | u_i \sim \mathrm{Ga}(\alpha, \beta/u_i),$$

independently for $i = 1, \dots, m$, where δ is a vector of unknown regression coefficients. In the above model, the random effect λ_i can be interpreted as an adjustment risk factor per unit area that is not explained by the auxiliary information. For λ_i, the global–local shrinkage prior is used to take account of the existence of singular towns whose crime risks are extremely high. Thus, the EH prior is used for u_i. For comparison, the standard gamma prior for λ_i, equivalent to setting $u_i = 1$ in the model, is applied. The results of posterior means of λ_i indicated that the use of the standard gamma prior produces estimates that seem to over-shrink the observed count and fails to detect extreme towns. On the other hand, the global–local shrinkage prior can detect such extreme towns due to the tail-robustness property. See Hamura et al. (2022) for more details and spatial mapping of the estimated crime risk.

8.2 Measurement Errors in Covariates

8.2.1 *Measurement Errors in the Fay–Herriot Model*

In the Fay–Herriot model (4.1), covariate or auxiliary information $\boldsymbol{x}_i$ is treated as a fixed value. However, $\boldsymbol{x}_i$ could be estimated from another survey in practice, thereby $\boldsymbol{x}_i$ might include estimation error, that is, $\boldsymbol{x}_i$ is measured with errors. Specifically, let $\widehat{\boldsymbol{x}}_i$ be the estimator of the true covariate $\boldsymbol{x}_i$, and assume that we observe $\widehat{\boldsymbol{x}}_i$ rather than $\boldsymbol{x}_i$, while the model (4.1) is defined with the true covariate $\boldsymbol{x}_i$. Further, suppose that its MSE matrix $\mathrm{MSE}(\widehat{\boldsymbol{x}}_i) = \boldsymbol{C}_i$ is known, like the sampling variance D_i in the Fay–Herriot model. In what follows, we also assume that $\widehat{\boldsymbol{x}}_i$ is independent of v_i and ε_i.

Ybarra and Lohr (2008) demonstrated the drawback of using the observed covariate $\widehat{\boldsymbol{x}}_i$ as if it were fixed. Remember that the BLUP of the small area mean $\theta_i = \boldsymbol{x}_i^\top \boldsymbol{\beta} + v_i$ is $\widetilde{\theta}_i = y_i - \gamma_i(y_i - \boldsymbol{x}_i^\top \boldsymbol{\beta})$ with $\gamma_i = D_i/(A + D_i)$ and the true covariate $\boldsymbol{x}_i$. Since we do not observe x_i, one may use an alternative naive predictor

$$\widetilde{\theta}_i^N = y_i - \gamma_i(y_i - \widehat{\boldsymbol{x}}_i^\top \boldsymbol{\beta}).$$

Note that $\widetilde{\theta}_i^N - \widetilde{\theta}_i = \gamma_i(\widehat{\boldsymbol{x}}_i - \boldsymbol{x}_i)^\top\boldsymbol{\beta}$. Then, the MSE of $\widetilde{\theta}_i^N$ can be evaluated as

$$\begin{aligned}\mathrm{E}[(\widetilde{\theta}_i^N - \theta_i)^2] &= \mathrm{E}[(\widetilde{\theta}_i - \theta_i)^2] + \mathrm{E}[(\widetilde{\theta}_i^N - \widetilde{\theta}_i)^2] \\ &= A\gamma_i + \gamma_i^2\boldsymbol{\beta}^\top\boldsymbol{C}_i\boldsymbol{\beta}.\end{aligned}$$

The second term $\gamma_i^2\boldsymbol{\beta}^\top\boldsymbol{C}_i\boldsymbol{\beta}$ corresponds to the MSE inflation due to using $\widehat{\boldsymbol{x}}_i$. Furthermore, MSE of the direct estimator y_i is $\mathrm{E}[(y_i - \theta_i)^2] = D_i$ so that MSE of the naive predictor $\widetilde{\theta}_i^N$ is larger than that of y_i when

$$A\gamma_i + \gamma_i^2\boldsymbol{\beta}^\top\boldsymbol{C}_i\boldsymbol{\beta} > D_i \quad \Leftrightarrow \quad \boldsymbol{\beta}^\top\boldsymbol{C}_i\boldsymbol{\beta} > A + D_i.$$

This means that if we use a noisy covariate $\widehat{\boldsymbol{x}}_i$, the resulting naive BLUP makes the direct estimator y_i worse instead of improving it.

To overcome this problem, we need to reconsider the form of BLUP. To this end, consider a class of predictors of the form $\widetilde{\mu}_i \equiv y_i - a_i(y_i - \widehat{\boldsymbol{x}}_i^\top\boldsymbol{\beta})$. Then, MSE of μ_i is given by

$$\begin{aligned}\mathrm{E}\left[(\widetilde{\mu}_i - \theta_i)^2\right] &= \mathrm{E}\left[\left\{(1-a_i)\varepsilon_i - a_i v_i + a_i(\widehat{\boldsymbol{x}}_i - \boldsymbol{x}_i)^\top\boldsymbol{\beta}\right\}^2\right] \\ &= (1-a_i)^2 D_i + a_i^2 A + a_i^2\boldsymbol{\beta}^\top\boldsymbol{C}_i\boldsymbol{\beta}.\end{aligned}$$

The above MSE is minimized at

$$a_i = \gamma_i^{\mathrm{ME}} \equiv \frac{D_i}{A + D_i + \boldsymbol{\beta}^\top\boldsymbol{C}_i\boldsymbol{\beta}},$$

and the minimum MSE is

$$\mathrm{E}\left[(\widetilde{\mu}_i - \theta_i)^2\right]\bigg|_{a_i=\gamma_i^{\mathrm{ME}}} = \frac{D_i(A + \boldsymbol{\beta}^\top\boldsymbol{C}_i\boldsymbol{\beta})}{A + D_i + \boldsymbol{\beta}^\top\boldsymbol{C}_i\boldsymbol{\beta}} < D_i.$$

Hence, the revised BLUP, defined as $\widetilde{\theta}_i^{\mathrm{ME}} = y_i - \gamma_i^{\mathrm{ME}}(y_i - \boldsymbol{x}_i^\top\boldsymbol{\beta})$ has a different shrinkage factor and always has smaller MSE than the direct estimator. Moreover, when $\boldsymbol{C}_i = 0$ (i.e., no measurement errors), $\widetilde{\theta}_i^{\mathrm{ME}}$ reduces to the standard BLUP.

For the parameter estimation, Ybarra and Lohr (2008) adopted the following general estimating equation for $\boldsymbol{\beta}$:

$$\sum_{i=1}^m w_i\left(\widehat{\boldsymbol{x}}_i\widehat{\boldsymbol{x}}_i^\top - \boldsymbol{C}_i\right)\boldsymbol{\beta} = \sum_{i=1}^m w_i\widehat{\boldsymbol{x}}_i y_i,$$

where $w_1, \ldots, w_m$ are fixed weights. The resulting estimator is

$$\widehat{\boldsymbol{\beta}}_w = \left\{ \sum_{i=1}^m w_i \left(\widehat{\boldsymbol{x}}_i \widehat{\boldsymbol{x}}_i^\top - \boldsymbol{C}_i \right) \right\}^{-1} \sum_{i=1}^m w_i \widehat{\boldsymbol{x}}_i y_i$$

if the inverse exists. Moreover, the random effect variance is estimated by

$$\widehat{A}_w = \frac{1}{m-p} \sum_{i=1} \left\{ \left(y_i - \widehat{\boldsymbol{x}}_i{}^\top \widehat{\boldsymbol{\beta}}_w \right)^2 - D_i - \widehat{\boldsymbol{\beta}}_w^\top \boldsymbol{C}_i \widehat{\boldsymbol{\beta}}_w \right\}.$$

In Ybarra and Lohr (2008), it is shown that both $\widehat{\boldsymbol{\beta}}_w$ and $\widehat{A}_w$ is consistent under some regularity conditions (e.g., finite $4 + \delta$ moments of $\widehat{\boldsymbol{x}}_i$, v_i and ε_i and uniform boundedness of w_i and D_i). Regarding the choice of w_i, Ybarra and Lohr (2008) recommended $w_i = (A + D_i + \boldsymbol{\beta}^\top \boldsymbol{C}_i \boldsymbol{\beta})^{-1}$, mimicking the generalized least squares estimator in the standard Fay–Herriot model. In this case, we initially set $w_i = 1$ and compute $\widehat{\boldsymbol{\beta}}_w$ and $\widehat{A}_w$. Then, we compute $\widehat{w}_i = (\widehat{A}_w + D_i + \widehat{\boldsymbol{\beta}}_w^\top \boldsymbol{C}_i \widehat{\boldsymbol{\beta}}_w)^{-1}$ to obtain $\widehat{\boldsymbol{\beta}}_{\widehat{w}}$ and $\widehat{A}_{\widehat{w}}$. This process can be iterated until convergence if desired. Using the parameter estimates, one can obtain the EBLUP of θ_i.

Arima et al. (2015) pointed out that the parameter estimation in the Fay–Herriot model with measurement errors could be unstable, and developed a hierarchical Bayesian approach to fit the model. The measurement error model is

$$y_i | \theta_i \sim \mathrm{N}(\theta_i, D_i), \quad \theta_i \sim \mathrm{N}(\boldsymbol{x}_i^\top \boldsymbol{\beta}, A), \quad \widehat{\boldsymbol{x}}_i | \boldsymbol{x}_i \sim \mathrm{N}(\boldsymbol{x}_i, \boldsymbol{C}_i), \quad i = 1, \ldots, m,$$

where D_i and $\boldsymbol{C}_i$ are known quantities. The prior distribution is

$$\pi(\boldsymbol{x}_1, \ldots, \boldsymbol{x}_m, \boldsymbol{\beta}, A) \propto 1.$$

The joint posterior distribution is

$$\begin{aligned} &\pi\left(\boldsymbol{\theta}, \boldsymbol{X}, \boldsymbol{\beta}, A \mid \boldsymbol{y}, \widehat{\boldsymbol{X}}\right) \\ &\propto A^{-m/2} \prod_{i=1}^m \exp\left[- \left\{ \frac{(y_i - \theta_i)^2}{2D_i} + \frac{\left(\theta_i - \boldsymbol{x}_i^\top \boldsymbol{\beta}\right)^2}{2A} + \frac{(\widehat{\boldsymbol{x}}_i - \boldsymbol{x}_i)^\top C_i^{-1} (\widehat{\boldsymbol{x}}_i - \boldsymbol{x}_i)}{2} \right\} \right], \end{aligned}$$

where $\boldsymbol{\theta} = (\theta_1, \ldots, \theta_m)$, $\boldsymbol{X} = (\boldsymbol{x}_1, \ldots, x_m)$ and $\boldsymbol{y} = (y_1, \ldots, y_m)$. The posterior distribution is proper when $m > p + 2$ and the posterior variances of $\boldsymbol{\beta}$ and A are finite when $m > p + 6$, where p is the dimension of $\boldsymbol{x}_i$. The posterior distribution cannot be obtained in closed form, but the implementation is greatly facilitated by the Gibbs sampler. The detailed sampling steps are given as follows:

$$\theta_i \mid \beta, A, y_i, \boldsymbol{x}_i \sim N\left(y_i - \frac{D_i}{A + D_i}(y_i - \boldsymbol{x}_i^\top \boldsymbol{\beta}), \frac{AD_i}{A + D_i}\right),$$

$$\boldsymbol{x}_i \mid \boldsymbol{\beta}, A, y_i, \widehat{\boldsymbol{x}}_i \sim N\left(\widehat{\boldsymbol{x}}_i + \frac{y_i - \widehat{\boldsymbol{x}}_i^\top \boldsymbol{\beta}}{A + D_i + \boldsymbol{\beta}^\top \boldsymbol{C}_i \boldsymbol{\beta}} \boldsymbol{C}_i \boldsymbol{\beta}, \ \boldsymbol{C}_i - \frac{\boldsymbol{C}_i \boldsymbol{\beta}\boldsymbol{\beta}^\top \boldsymbol{C}_i}{A + D_i + \boldsymbol{\beta}^\top \boldsymbol{C}_i \boldsymbol{\beta}}\right),$$

$$\boldsymbol{\beta} \mid A, \boldsymbol{\theta}, \boldsymbol{X} \sim N\left(\left(\boldsymbol{X}^\top \boldsymbol{X}\right)^{-1} \boldsymbol{X}^\top \boldsymbol{\theta}, A\left(\boldsymbol{X}^\top \boldsymbol{X}\right)^{-1}\right),$$

$$A \mid \boldsymbol{\beta}, \boldsymbol{\theta}, \boldsymbol{X} \sim \mathrm{IG}\left(\frac{1}{2}(m-2), \frac{1}{2}\sum_{i=1}^{m}\left(\theta_i - \boldsymbol{x}_i^\top \boldsymbol{\beta}\right)^2\right).$$

Based on the posterior samples of θ_i, we may use posterior mean and variance as a point estimate of θ_i and uncertainty measure of the estimate, respectively.

8.2.2 Measurement Errors in the Nested Error Regression Model

Measurement error problems can also occur in the nested error regression model. Ghosh et al. (2006) and Torabi et al. (2009) considered the following nested error regression model with a structural measurement error:

$$\begin{aligned} y_{ij} &= \beta_0 + \beta_1 x_i + v_i + \varepsilon_{ij}, \quad v_i \sim \mathrm{N}(0, \tau^2), \quad \varepsilon_{ij} \sim \mathrm{N}(0, \sigma_\varepsilon^2), \\ X_{ij} &= x_i + \eta_{ij}, \quad \eta_{ij} \sim \mathrm{N}(0, \sigma_\eta^2), \quad x_i \sim \mathrm{N}(\mu, \sigma_x^2), \end{aligned}$$

where v_i, ε_{ij} and η_{ij} are mutually independent. Here, x_{ij} is the true covariate and X_{ij} is the observed covariate measured with errors. The vector of model parameters are $\boldsymbol{\phi} = (\beta_0, \beta_1, \mu, \tau^2, \sigma_\varepsilon^2, \sigma_\eta^2, \sigma_x^2)$. Suppose that we are interested in the finite population mean, $\theta_i = N_i^{-1}\sum_{j=1}^{N_i} y_{ij}$, and we only observe $\boldsymbol{y}_i^{(s)} = (y_{i1}, \ldots, y_{in_i})$ and $\boldsymbol{X}_i^{(s)} = (\boldsymbol{x}_{i1}, \ldots, \boldsymbol{x}_{in_i})$. Then, the conditional distribution of $\boldsymbol{y}_i^{(r)}$ given $\boldsymbol{y}_i^{(s)}$ is the multivariate normal distribution with mean vector and covariance matrix given by

$$\begin{aligned} &\mathrm{E}\left(\boldsymbol{y}_i^{(r)} \mid \boldsymbol{y}_i^{(s)}, \boldsymbol{X}_i^{(s)}, \boldsymbol{\phi}\right) \\ &= \left[(1 - A_i)\bar{y}_i + A_i(\beta_0 + \beta_1 \mu_x) + A_i\left\{\frac{n_i \sigma_x^2}{\sigma_\eta^2 + n_i \sigma_x^2}\right\}\beta_1\left(\bar{X}_i - \mu_x\right)\right]\boldsymbol{1}_{N_i - n_i} \end{aligned}$$

and

$$\mathrm{Var}\left(\boldsymbol{y}_i^{(r)} \mid \boldsymbol{y}_i^{(s)}, \boldsymbol{X}_i^{(s)}, \boldsymbol{\phi}\right) = \sigma_\varepsilon^2 \boldsymbol{I}_{N_i - n_i} + A_i\left\{\beta_1^2 \sigma_x^2 + \tau^2 - \frac{n_i \beta_1^2 \sigma_x^4}{\sigma_\eta^2 + n_i \sigma_x^2}\right\}\boldsymbol{J}_{N_i - n_i},$$

where $\mathbf{1}_{N_i-n_i}$ is the $(N_i - n_i) \times 1$ vector of 1's, $\mathbf{J}_{N_i-n_i} = \mathbf{1}_{N_i-n_i}\mathbf{1}_{N_i-n_i}^\top$, $\overline{X}_i = n_i^{-1}\sum_{j=1}^{n} X_{ij}$, and

$$A_i = \frac{\sigma_\varepsilon^2\left(\sigma_\eta^2 + n_i\sigma_x^2\right)}{n_i\beta_1^2\sigma_x^2\sigma_\eta^2 + \left(n_i\tau^2 + \sigma_\varepsilon^2\right)\left(\sigma_\eta^2 + n_i\sigma_x^2\right)}.$$

Then, the best predictor of θ_i is

$$\widetilde{\theta}_i(\boldsymbol{\phi}) = \frac{1}{N_i}\left\{\sum_{j=1}^{n_i} y_{ij} + \mathbf{1}_{N_i-n_i}^\top \mathrm{E}\left(\boldsymbol{y}_i^{(r)} \mid \boldsymbol{y}_i^{(s)}, \boldsymbol{X}_i^{(s)}, \boldsymbol{\phi}\right)\right\}.$$

One can estimate $\boldsymbol{\phi}$ via the moment method. We first estimate σ_η^2 and σ_ε^2 as

$$\widehat{\sigma}_\eta^2 = \frac{1}{n_T - m}\sum_{i=1}^{m}\sum_{j=1}^{n_i}(X_{ij} - \overline{X}_i)^2, \quad \widehat{\sigma}_\varepsilon^2 = \frac{1}{n_T - m}\sum_{i=1}^{m}\sum_{j=1}^{n_i}(y_{ij} - \overline{y}_i)^2,$$

and then estimate the other parameters as

$$\widehat{\beta}_1 = \frac{\sum_{i=1}^m n_i\overline{y}_i\left(\overline{X}_i - \overline{X}\right)}{\sum_{i=1}^m n_i\left(\overline{X}_i - \overline{X}\right)^2 - (m-1)\widehat{\sigma}_\eta^2}, \quad \widehat{\beta}_0 = \overline{y} - \widehat{\beta}_1\overline{X}, \quad \widehat{\mu} = \overline{X},$$

$$\widehat{\sigma}_x^2 = \frac{1}{g_m}\left(\sum_{i=1}^{m} n_i\left(\overline{X}_i - \overline{X}\right)^2 - (m-1)\widehat{\sigma}_\eta^2\right),$$

$$\widehat{\tau}^2 = \frac{1}{g_m}\left(\sum_{i=1}^{m} n_i(\overline{y}_i - \overline{y})^2 - (m-1)\widehat{\sigma}_\varepsilon^2\right) - \widehat{\beta}_1^2\widehat{\sigma}_x^2,$$

where $\overline{X} = n_T^{-1}\sum_{i=1}^m n_i\overline{X}_i$, $\overline{y} = n_T^{-1}\sum_{i=1}^m n_i\overline{y}_i$, $g_m = n_T - \sum_i n_i^2/n_T$, and $n_T = \sum_{i=1}^m n_i$. Since the estimators $\widehat{\sigma}_x^2$ and τ^2 may produce negative values, we should use the positive part estimators, $\widehat{\sigma}_{x,P}^2 = \max(0, \widehat{\sigma}_x^2)$ and $\widehat{\tau}_{x,P}^2 = \max(0, \widehat{\tau}^2)$. The empirical best predictor of θ_i is obtained as $\widehat{\theta}_i = \widetilde{\theta}_i(\widehat{\boldsymbol{\phi}})$.

8.3 Nonparametric and Semiparametric Modeling

In the standard mixed models in small area estimation, parametric models are typically adopted for simplicity. However, parametric models suffer from model misspecification, which could produce unreliable small area estimates. Instead of using parametric models, the use of nonparametric or semiparametric models has been considered in the literature.

Opsomer et al. (2008) proposed nonparametric estimation of the regression part in the linear mixed model by adopting the P-spline method. Specifically, the authors proposed the nested error regression model with the nonparametric mean term, written as

$$y_{ij} = f(x_{ij}) + v_i + \varepsilon_{ij}, \quad j = 1, \dots, n_i, \quad i = 1, \dots, m, \tag{8.6}$$

where $v_i \sim \mathrm{N}(0, \tau^2)$ and $\varepsilon_{ij} \sim \mathrm{N}(0, \sigma^2)$, and $f(\cdot)$ is an unknown mean function. For simplicity, we first consider a single covariate x_{ij}, but the extension to multiple covariates will be discussed later. Here, $f(\cdot)$ is modeled by the P-spline of the form

$$f(x; \boldsymbol{\beta}, \boldsymbol{\gamma}) = \beta_0 + \beta_1 x + \cdots + \beta_p x^p + \sum_{\ell=1}^{K} \gamma_\ell (x - \kappa_\ell)_+^p,$$

where p is the degree of the spline, $(x)_+^p = x^p I(x > 0)$, $\kappa_1 < \ldots < \kappa_K$ is a set of fixed knots and $\boldsymbol{\beta} = (\beta_0, \dots, \beta_q)$ and $\boldsymbol{\gamma} = (\gamma_1, \dots, \gamma_K)$ denote coefficient vectors for the parametric and spline terms. Provided that the knot locations are sufficiently spread out over the range of x and K is sufficiently large, the class of function $f(x; \boldsymbol{\beta}, \boldsymbol{\gamma})$ can approximate most smooth functions. The knots are often at equally spaced quantiles of the covariate, and K is taken to be large relative to the size. A typical knot choice for univariate covariate would be one knot every four or five observations.

To prevent over-fitting, the ridge penalization is typically introduced for $\boldsymbol{\gamma}$, which is equivalent to introducing a prior distribution, $\boldsymbol{\gamma} \sim \mathrm{N}(0, \lambda \boldsymbol{I}_K)$, where λ is an unknown parameter controlling the smoothness of the estimation of $f(\cdot)$. Let $\boldsymbol{z}_{ij} = (1, x_{ij}, \dots, x_{ij}^p)$ and $\boldsymbol{w}_{ij} = ((x_{ij} - \kappa_1)_+^p), \dots, (x_{ij} - \kappa_K)_+^p)$. Then the nonparametric model (8.6) can be expressed as a linear mixed model given by

$$y_{ij} = \boldsymbol{z}_{ij}^\top \boldsymbol{\beta} + \boldsymbol{w}_{ij}^\top \boldsymbol{\gamma} + v_i + \varepsilon_{ij}, \tag{8.7}$$

where $\boldsymbol{\gamma} \sim \mathrm{N}(0, \lambda \boldsymbol{I}_K)$ and $v_i \sim \mathrm{N}(0, \tau^2)$. Define $\boldsymbol{Z}_i = (\boldsymbol{z}_{i1}, \dots, \boldsymbol{z}_{in_i})$, $\boldsymbol{Z} = (\boldsymbol{Z}_1, \dots, \boldsymbol{Z}_m)^\top$, $\boldsymbol{W}_i = (\boldsymbol{w}_{i1}, \dots, \boldsymbol{w}_{in_i})$, $\boldsymbol{W} = (\boldsymbol{W}_1, \dots, \boldsymbol{W}_m)^\top$, and $\boldsymbol{D} = \mathrm{blockdiag}(\boldsymbol{1}_{n_1}, \dots, \boldsymbol{1}_{n_m})$. Then, the model (8.7) can be rewritten as

$$\boldsymbol{Y} = \boldsymbol{Z\beta} + \boldsymbol{W\gamma} + \boldsymbol{Dv} + \boldsymbol{\epsilon},$$

where $\boldsymbol{Y} = (y_{11}, \dots, y_{1n_1}, y_{21}, \dots, y_{mn_m})^\top$, $\boldsymbol{v} = (v_1, \dots, v_m)^\top$, and $\boldsymbol{\epsilon}$ is defined in the same way as $\boldsymbol{y}$. Thus, the nonparametric model (8.6) with the P-spline can be written as a linear mixed model with two random effects, $\boldsymbol{\gamma}$ and $\boldsymbol{v}$. Note that $\mathrm{Var}(\boldsymbol{Y}) = \boldsymbol{V} \equiv \lambda \boldsymbol{W}\boldsymbol{W}^\top + \tau^2 \boldsymbol{D}\boldsymbol{D}^\top + \sigma^2 \boldsymbol{I}_N$ with $N = \sum_{i=1}^m ni$. The standard theory of BLUP shows that the generalized least squares estimator of $\boldsymbol{\beta}$ is

$$\widetilde{\boldsymbol{\beta}} = (\boldsymbol{Z}^\top \boldsymbol{V}^{-1} \boldsymbol{Z})^{-1} \boldsymbol{X}^\top \boldsymbol{V}^{-1} \boldsymbol{Y}$$

and the predictors of $\boldsymbol{\gamma}$ and $\boldsymbol{v}$ are

$$\tilde{\boldsymbol{\gamma}} = \lambda \boldsymbol{W}^{\mathrm{T}} \boldsymbol{V}^{-1} \left(\boldsymbol{Y} - \boldsymbol{X}\tilde{\boldsymbol{\beta}} \right), \quad \tilde{\boldsymbol{v}} = \tau^2 \boldsymbol{D}^{\mathrm{T}} \boldsymbol{V}^{-1} \left(\boldsymbol{Y} - \boldsymbol{X}\tilde{\boldsymbol{\beta}} \right).$$

If we are interested in $\mu_i = f(c_i) + v_i$ for some fixed quantity c_i, we can predict μ_i by $\widetilde{\mu}_i = \widetilde{z}_i^{\top} \widetilde{\boldsymbol{\beta}} + \widetilde{w}_i^{\top} \widetilde{\boldsymbol{\gamma}} + \widetilde{v}_i$, where $\widetilde{z}_i = (1, c_i, \dots, c_i^p)^{\top}$ and $\widetilde{w}_i = ((c_i - \kappa_1)_+^p, \dots, (c_i - \kappa_K)_+^p)^{\top}$. Regarding the estimation of unknown variance parameters, λ, τ^2 and σ^2, we may use (residual) maximum likelihood methods.

When there are multiple covariates, the P-spline formulation can be extended to the multidimensional case by taking tensor products of univariate basis functions. However, this strategy leads to a considerably large number of basis functions when the number of covariates is not small. Instead, it would be convenient to use radial basis functions. Suppose $\boldsymbol{x}$ is a q-dimensional covariate vector. The Gaussian radial basis functions are defined as $C_\ell(\boldsymbol{x}) = \exp(-a \,|\, \boldsymbol{x} - \boldsymbol{\kappa}_\ell \,|^2)$ $(\ell = 1, \dots, K)$, where $\boldsymbol{\kappa}_1, \dots, \boldsymbol{\kappa}_K \in \mathbb{R}^q$ are predetermined knots and $a > 0$ is a tuning parameter controlling the decay of the radial basis. Then, $f(\boldsymbol{x}_{ij})$ with $\boldsymbol{x}_{ij} \in \mathbb{R}^q$ can be modeled as $\boldsymbol{w}_{ij}^{\top} \boldsymbol{\gamma}$ with $\boldsymbol{w}_{ij} = (C_1(\boldsymbol{x}_{ij}), \dots, C_K(\boldsymbol{x}_{ij}))^{\top}$.

8.4 Modeling Heteroscedastic Variance

8.4.1 Shrinkage Estimation of Sampling Variances

One of the main criticisms of the FH model (4.1) is the assumption that the sampling variances D_i are known although they are actually estimated/computed from data. It has been revealed that the assumption for D_i may lead to several serious problems such as underestimation of risks (Wang and Fuller 2003). In order to take account of variability of D_i, You and Chapman (2006) introduced the following joint hierarchical model for y_i and D_i:

$$\begin{aligned} &y_i | \theta_i, \sigma_i^2 \sim \mathrm{N}(\theta_i, \sigma_i^2), \quad \theta_i \sim \mathrm{N}(\boldsymbol{x}_i^{\top} \boldsymbol{\beta}, A), \\ &D_i | \sigma_i^2 \sim \mathrm{Ga}\left(\frac{n_i - 1}{2}, \frac{n_i - 1}{2\sigma_i^2} \right), \quad \sigma_i^2 \sim \pi(\sigma_i^2), \end{aligned} \tag{8.8}$$

where n_i is a sample size, and $\pi(\cdot)$ is a prior distribution for σ_i^2. In the model (8.8), θ_i and σ_i^2 are the true mean and variance, and y_i and D_i are estimates of them, respectively. You and Chapman (2006) adopted $\mathrm{Ga}(a_i, b_i)$ for $\pi(\cdot)$, where a_i and b_i are fixed constants, and recommended using small values for a_i and b_i, leading to diffuse priors for σ_i^2. This means that the resulting estimator of σ_i^2 does not hold shrinkage effect such that the posterior mean of σ_i^2 is almost the same as D_i. On the other hand, Dass et al. (2012) and Maiti et al. (2014) adopted the model

$\sigma_i^2 \sim \mathrm{Ga}(\alpha, \gamma)$, where α and γ are unknown parameters. However, the estimation of these parameters via an EM algorithm tends to be unstable. To overcome the drawback, Sugasawa et al. (2017) proposed an alternative model

$$\sigma_i^2 \sim \mathrm{Ga}(a_i, b_i\gamma), \quad i = 1, \dots, m, \tag{8.9}$$

with unknown γ and fixed constants of a_i and b_i. Sugasawa et al. (2017) developed a hierarchical Bayesian approach using the model (8.8) with the above model for σ_i^2, where the prior distribution is $\pi(\boldsymbol{\beta}, A, \gamma) \propto 1$. Then, the joint posterior distribution is

$$\pi\left(\boldsymbol{\theta}, \boldsymbol{\sigma}^2, \boldsymbol{\beta}, A, \gamma \mid \boldsymbol{y}, \boldsymbol{D}\right) \propto A^{-m/2} \prod_{i=1}^m \gamma^{a_i} \left(\sigma_i^2\right)^{-n_i/2-a_i-1} \times \exp\left\{-\frac{(y_i-\theta_i)^2 + (n_i-1)D_i^2 + 2b_i\gamma}{2\sigma_i^2} - \frac{\left(\theta_i - \boldsymbol{x}_i^\top\boldsymbol{\beta}\right)^2}{2A}\right\},$$

where $\boldsymbol{\theta} = (\theta_1, \dots, \theta_m)$, $\boldsymbol{\sigma}^2 = (\sigma_1^2, \dots, \sigma_m^2)$, $\boldsymbol{y} = (y_1, \dots, y_m)$ and $\boldsymbol{D} = (D_1, \dots, D_m)$. The following two properties can be shown.

- The marginal posterior density $\pi\left(\boldsymbol{\beta}, \tau^2, \gamma \mid \boldsymbol{y}, \boldsymbol{D}\right)$ is proper if $m > p+2$, $n_i > 1$ and $\mathrm{rank}(\boldsymbol{X}) = p$, where $\boldsymbol{X} = (\boldsymbol{x}_1, \dots, \boldsymbol{x}_m)$.
- The model parameters $\boldsymbol{\beta}$, τ^2 and γ have finite posterior variances if $m > p + 6$, $n_i > 1$ and $\mathrm{rank}(\boldsymbol{X}) = p$.

The above two properties guarantee the use of the posterior means and variances to summarize the posterior distribution. Although the posterior distributions cannot be obtained in an analytical way, one can generate posterior samples by a Gibbs sampler with full conditional distributions given by

$$\theta_i \mid \boldsymbol{\beta}, A, \sigma_i^2, y_i \sim N\left(y_i - \frac{\sigma_i^2}{A+\sigma_i^2}(y_i - \boldsymbol{x}_i^\top\boldsymbol{\beta}), \frac{A\sigma_i^2}{A+\sigma_i^2}\right), \quad i = 1, \dots, m,$$

$$\sigma_i^2 \mid \gamma, \theta_i, y_i, D_i \sim \mathrm{IG}\left(\frac{n_i}{2} + a_i, \frac{1}{2}(y_i-\theta_i)^2 + \frac{1}{2}(n_i-1)D_i^2 + b_i\gamma\right), \quad i = 1, \dots, m$$

$$A \mid \boldsymbol{\beta}, \boldsymbol{\theta} \sim \mathrm{IG}\left(\frac{m}{2} - 1, \frac{1}{2}(\boldsymbol{\theta} - \boldsymbol{X}\boldsymbol{\beta})^\top(\boldsymbol{\theta} - \boldsymbol{X}\boldsymbol{\beta})\right), \quad \gamma \mid \boldsymbol{\sigma}^2 \sim \Gamma\left(\sum_{i=1}^m a_i + 1, \sum_{i=1}^m \frac{b_i}{\sigma_i^2}\right),$$

$$\boldsymbol{\beta} \mid A, \boldsymbol{\theta} \sim N_p\left(\left(\boldsymbol{X}^\top\boldsymbol{X}\right)^{-1}\boldsymbol{X}^\top\boldsymbol{\theta}, A\left(\boldsymbol{X}^\top\boldsymbol{X}\right)^{-1}\right).$$

Since all the full conditional distributions are familiar ones, one can easily generate posterior samples.

Regarding the choice of a_i and b_i, we first observe that

$$\mathrm{Var}(y_i) = \mathrm{E}[\mathrm{Var}(y_i \mid \theta_i)] + \mathrm{Var}(\mathrm{E}[y_i \mid \theta_i]) = \mathrm{E}\left[\sigma_i^2\right] = \frac{b_i}{a_i - 1}\gamma.$$

Since y_i is the sample mean, it is reasonable to assume that $\mathrm{Var}(y_i) = O\left(n_i^{-1}\right)$. From the full conditional expectation of σ_i^2, it follows that

$$\begin{aligned}\mathrm{E}\left[\sigma_i^2 \mid \gamma, \theta_i, y_i, D_i\right] &= \frac{(y_i - \theta_i)^2/2 + (n_i - 1) D_i/2 + b_i \gamma}{n_i/2 + a_i - 1} \\ &= \frac{n_i/2}{n_i/2 + a_i - 1} \widetilde{\sigma}_i^2 (y_i, D_i) + \frac{a_i - 1}{n_i/2 + a_i - 1} \cdot \frac{b_i}{a_i - 1} \gamma,\end{aligned}$$

where

$$\widetilde{\sigma}_i^2 (y_i, D_i) = \frac{1}{n_i} \left\{(y_i - \theta_i)^2 + (n_i - 1) D_i\right\}.$$

It is observed that the full conditional expectation of σ_i^2 is the weighted mean of $\widetilde{\sigma}_i^2 (y_i, D_i)$ and the prior mean $b_i \gamma / (a_i - 1)$, and the weight for the prior mean is determined by a_i. Since $\mathrm{E}\left[\sigma_i^2 \mid \gamma, \theta_i, y_i, D_i\right]$ approaches to D_i for large n_i, it would be natural to set a_i and b_i such that $a_i = O(1)$ and $b_i = O\left(n_i^{-1}\right)$. Specifically, Sugasawa et al. (2017) suggested $a_i = 2$ and $b_i = n_i^{-1}$. Sugasawa et al. (2017) also proposed an alternative modeling for the heteroscedastic variance using some covariates, given by

$$\sigma_i^2 \sim IG\left(a_i, b_i \gamma \exp\left(z_i^\top \eta\right)\right), \tag{8.10}$$

where z_i is a vector of covariate and η is an unknown vector. This model can assist in the modeling of σ_i^2 via covariate information, which may improve the estimation accuracy of σ_i^2. Unlike the model (8.9), the full conditional distribution of η under the model (8.10) is not a familiar form. Thus we use the random-walk Metropolis–Hastings algorithm to generate samples of η.

Example 8.4 (*Simulation study*) Sugasawa et al. (2017) conducted simulation studies to compare the estimators of hierarchical and empirical Bayes methods with estimated D_i's. To generate synthetic data, unit observations in each area are generated from

$$Y_{ij} = \beta_0 + \beta_1 x_i + v_i + \varepsilon_{ij}, \quad j = 1, \ldots, n_i, \quad i = 1, \ldots, m,$$

where $m = 30$, $n_i = 7$, $v_i \sim \mathrm{N}(0, A)$ and $\varepsilon_{ij} \sim \mathrm{N}(n_i \sigma_i^2)$. Then, the model for the sample mean $y_i = \bar{Y}_i \equiv n_i^{-1} \sum_{j=1}^{n_i} Y_{ij}$ is

$$y_i \sim \mathrm{N}(\theta_i, \sigma_i^2), \quad \theta_i \sim \mathrm{N}(\beta_0 + \beta_1 x_i, A).$$

The direct estimator of σ_i^2 is defined as $D_i = n_i^{-1}(n_i - 1)^{-1} \sum_{j=1}^{n_i} (Y_{ij} - \bar{Y}_i)^2$, so that $D_i | \sigma_i^2 \sim \mathrm{Ga}((n_i - 1)/2, (n_i - 1)/2\sigma_i^2)$. The covariate x_i is generated from the uniform distribution on (2, 8) and the true parameter values are $\beta_0 = 0.5$, $\beta_1 = 0.8$, and $A = 1$. For the true values of σ_i^2, the following two scenarios are considered.

$$\text{(I)}\ \sigma_i^2 \sim \mathrm{IG}(10, 5\exp(0.3x_i)), \qquad \text{(II)}\ \sigma_i^2 \sim U(0.5, 5).$$

For the generated dataset, two hierarchical Bayesian methods with the models (8.9) and (8.10), denoted by STK1 and STK2, respectively, and the method by You and Chapman (2006), denoted by YC are applied. Moreover, the empirical Bayes method by Maiti et al. (2014), denoted by MRS, is applied. To measure the performance, MSE and bias of point estimates of θ_i and σ_i^2 are adopted. The results are summarized as follows:

- For θ_i, both STK1 and STK2 provide smaller MSE values than YC, while the bias of the three estimates is comparable. This indicates the importance of shrinkage estimation of σ_i^2 to improve the estimation accuracy of θ_i.
- For σ_i^2, the bias of YC is smaller than STK1 and STK2, and the MSE of YC is much larger (almost twice as large as) than those of STK1 and STK2. This is because the YC method does not provide shrinkage estimation of σ_i^2.
- MRS provides comparable MSE values for θ_i as STK1 and STK2, but it has a large bias. Also, MSE for σ_i^2 of MRS is almost twice as large as that of STK1 and STK2.
- STK2 slightly performs better than STK1 under scenario (I), since the covariate-dependent modeling (8.10) is effective in the scenario.

8.4.2 *Heteroscedastic Variance in Nested Error Regression Models*

In the NER model, the variance parameters in random effects and error terms are assumed to be constants in the classical NER model (4.19), but it could be unrealistic in some applications. To address this issue, Jiang and Nguyen (2012) proposed the following heteroscedastic nested error regression models:

$$y_i = \boldsymbol{x}_{ij}^\top \boldsymbol{\beta} + v_i + \varepsilon_{ij}, \quad v_i \sim \mathrm{N}(0, \lambda\sigma_i^2), \quad \varepsilon_{ij} \sim \mathrm{N}(0, \sigma_i^2), \tag{8.11}$$

where $\lambda, \sigma_1^2, \ldots, \sigma_m^2$ are unknown parameters. Since the number of $\sigma_1^2, \ldots, \sigma_m^2$ increases with the number of areas m, these parameters are typically called "incidental parameters". A crucial assumption here is that $\mathrm{Var}(v_i)/\mathrm{Var}(\varepsilon_{ij}) = \lambda$, that is, the ratio of random effect variance and error variance is constant. Under the model, the best predictor of v_i is given by

$$\widetilde{v}_i(\boldsymbol{\beta}, \lambda) = \frac{n_i\lambda\sigma_i^2}{\sigma_i^2 + n_i\lambda\sigma_i^2}\left(\bar{y}_i - \bar{\boldsymbol{x}}_i^\top\boldsymbol{\beta}\right) = \frac{n_i\lambda}{1 + n_i\lambda}\left(\bar{y}_i - \bar{\boldsymbol{x}}_i^\top\boldsymbol{\beta}\right),$$

where $\bar{y}_i = n_i^{-1}\sum_{j=1}^{n_i} y_{ij}$ and $\bar{\boldsymbol{x}}_i = n_i^{-1}\sum_{j=1}^{n_i}\boldsymbol{x}_{ij}$. Hence, the best predictor depends on the heteroscedastic variance only through the ratio $\mathrm{Var}(v_i)/\mathrm{Var}(\varepsilon_{ij})$, which does not depend on i.

The log-likelihood function for $\boldsymbol{\beta}$, λ and σ_i^2 has the expression

$$
\begin{aligned}
l(\boldsymbol{\beta}, \lambda, \boldsymbol{\sigma}^2) = C - \frac{1}{2}\sum_{i=1}^{m}\left[n_i \log\left(\sigma_i^2\right) + \log(1+n_i\lambda)\right] \\
+ \sum_{i=1}^{m}\left[\frac{1}{\sigma_i^2}\left\{\sum_{j=1}^{n_i}\left(y_{ij} - \boldsymbol{x}_{ij}^\top\boldsymbol{\beta}\right)^2 - \frac{\lambda}{1+n_i\lambda}\left(\bar{y}_i - \bar{\boldsymbol{x}}_i^\top\boldsymbol{\beta}\right)^2\right\}\right],
\end{aligned}
$$

where C does not depend on the parameters. Given β and λ, it is easy to obtain the maximum likelihood estimator of σ_i^2, that is,

$$
\widetilde{\sigma}_i^2(\boldsymbol{\beta}, \lambda) = \frac{1}{n_i}\left\{\sum_{j=1}^{n_i}\left(y_{ij} - \boldsymbol{x}_{ij}^\top\boldsymbol{\beta}\right)^2 - \frac{\lambda}{1+n_i\lambda}\left(\bar{y}_i - \bar{\boldsymbol{x}}_i^\top\boldsymbol{\beta}\right)^2\right\}.
$$

By replacing σ_i^2 with $\widetilde{\sigma}_i^2$ in the log-likelihood $l(\boldsymbol{\beta}, \lambda, \boldsymbol{\sigma}^2)$, we obtain the profile likelihood given by

$$
\widetilde{l}(\boldsymbol{\beta}, \lambda) \equiv l(\boldsymbol{\beta}, \lambda, \widetilde{\boldsymbol{\sigma}}^2(\boldsymbol{\beta}, \lambda)) = \widetilde{C} - \frac{1}{2}\sum_{i=1}^{m}\left\{n_i \log\left(\widetilde{\sigma}_i^2(\boldsymbol{\beta}, \lambda)\right) + \log(1+n_i\lambda)\right\},
$$

where $\widetilde{C}$ does not depend on the parameters. Thus, the maximum likelihood estimator of $(\boldsymbol{\beta}, \lambda)$ is defined as the maximizer of $\widetilde{l}(\boldsymbol{\beta}, \lambda)$. Under some regularity conditions, Jiang and Nguyen (2012) proved that the profile likelihood estimator of $\boldsymbol{\beta}$ and λ is consistent when $n_i > 1$ and $m \to \infty$. It should be noted that the estimator $\widetilde{\sigma}_i^2$ is not consistent as long as n_i is finite. Jiang and Nguyen (2012) also showed that the maximum likelihood estimator of $\lambda \equiv \tau^2/\sigma^2$ under the standard nested error regression (4.19) is typically inconsistent when the underlying variance is heteroscedastic.

Since the best predictor $\widetilde{v}_i$ does not depend on the incidental parameter σ_i^2, the empirical best predictor $\widehat{v}_i \equiv \widetilde{v}_i(\widehat{\boldsymbol{\beta}}, \widehat{\lambda})$ would perform well. On the other hand, the MSE of $\widehat{v}_i$ is expressed as

$$
\mathrm{E}[(\widehat{v}_i - v_i)^2] = \mathrm{E}[(\widetilde{v}_i - v_i)^2] + \mathrm{E}[(\widehat{v}_i - \widetilde{v}_i)^2]
$$

and

$$
\mathrm{E}[(\widetilde{v}_i - v_i)^2] = \frac{\lambda\sigma_i^2}{1+n_i\lambda}.
$$

Since the leading term of the MSE depends on σ_i^2, it is not possible to consistently estimate the MSE without any additional assumptions for σ_i^2. Jiang and Nguyen (2012) adopted the assumption that m areas can be divided into q groups, namely, $\{1, \ldots, m\} = S_1 \cup \cdots \cup S_q$, such that $\mathrm{E}[\sigma_i^2] = \phi_t$ for $i \in S_t$ and $t = 1, \ldots, q$. Under the assumption, it follows that $\mathrm{E}[(\widetilde{v}_i - v_i)^2] = \lambda\phi_t/(1+n_i\lambda)$ for $i \in S_t$, and ϕ_t can be consistently estimated as long as $|S_t| \to \infty$ for every t.

Kubokawa et al. (2016) proposed an alternative modeling of heteroscedastic variance named "random dispersion" models. In addition to the heteroscedastic model (8.11), it is assumed that $\eta_i \equiv 1/\sigma_i^2$ are mutually independent and identically distributed as

$$\eta_i \sim \mathrm{Ga}\left(\frac{\tau_1}{2}, \frac{2}{\tau_2}\right),$$

with unknown parameters τ_1 and τ_2. Note that $\mathrm{E}[\sigma_i^2] = \tau_2/(\tau_1 - 2)$ under the above gamma model. Let $\boldsymbol{\omega} = (\boldsymbol{\beta}^\top, \lambda, \tau_1, \tau_2)$ be a vector of unknown parameters. The marginal joint distribution of $\boldsymbol{y} = \left(\boldsymbol{y}_1, \ldots, \boldsymbol{y}_m\right)^\top$ and $\boldsymbol{\eta} = (\eta_1, \ldots, \eta_m)^\top$ after integrating v_i's out can be expressed as

$$f(\boldsymbol{y}, \boldsymbol{\eta} \mid \boldsymbol{\omega}) = \prod_{i=1}^{m} \left\{ \frac{\tau_2^{\tau_1/2} \eta_i^{(n_i+\tau_1)/2-1} 2^{-(n_i+\tau_1)/2}}{\pi^{n_i/2} \Gamma\left(\tau_1/2\right) \sqrt{n_i \lambda + 1}} \exp\left[-\frac{\eta_i}{2} \left\{ Q_i\left(\boldsymbol{y}_i, \boldsymbol{\beta}, \lambda\right) + \tau_2 \right\} \right] \right\},$$

where

$$Q_i\left(\boldsymbol{y}_i, \boldsymbol{\beta}, \lambda\right) = \sum_{j=1}^{n_i} \left\{ \left(y_{ij} - \bar{y}_i\right) - \left(\boldsymbol{x}_{ij} - \bar{\boldsymbol{x}}_i\right)^\top \boldsymbol{\beta} \right\}^2 + n_i \gamma_i(\lambda) \left(\bar{y}_i - \bar{\boldsymbol{x}}_i^\top \boldsymbol{\beta}\right)^2,$$

and $\gamma_i(\lambda) = 1/(n_i\lambda + 1)$. Integrating out the joint distribution with respect to $\boldsymbol{\eta}$, one can obtain the marginal distribution of $\boldsymbol{y}$ as

$$f(\boldsymbol{y} \mid \boldsymbol{\omega}) = \prod_{i=1}^{m} \left\{ \frac{\tau_2^{\tau_1/2} \Gamma\left(\left(n_i + \tau_1\right)/2\right)}{\pi^{n_i/2} \sqrt{n_i \lambda + 1} \Gamma\left(\tau_1/2\right)} \left\{ Q_i\left(\boldsymbol{y}_i, \boldsymbol{\beta}, \lambda\right) + \tau_2 \right\}^{-(n_i+\tau_1)/2} \right\}.$$

Then, the log-likelihood function of $\boldsymbol{\omega}$ can be obtained as

$$\begin{aligned} L(\boldsymbol{\omega}) = &-\sum_{i=1}^{m} n_i \log \pi + m\tau_1 \log \tau_2 + 2\sum_{i=1}^{m} \log\left\{ \Gamma\left(\frac{n_i + \tau_1}{2}\right) \right\} - 2\,m \log\left\{ \Gamma\left(\frac{\tau_1}{2}\right) \right\} \\ &- \sum_{i=1}^{m} \log\left(n_i \lambda + 1\right) - \sum_{i=1}^{m} \left(n_i + \tau_1\right) \log\left(Q_i + \tau_2\right). \end{aligned}$$

Then, the maximum likelihood estimator of $\boldsymbol{\omega}$ is $\widehat{\boldsymbol{\omega}} = \mathrm{argmax}_{\boldsymbol{\omega}} L(\boldsymbol{\omega})$. Kubokawa et al. (2016) derived the following asymptotic properties of $\widehat{\boldsymbol{\omega}} = (\widehat{\boldsymbol{\beta}}^\top, \widehat{\boldsymbol{\theta}}^\top)$:

$$\begin{aligned} \mathrm{E}\left[(\widehat{\boldsymbol{\beta}} - \boldsymbol{\beta})(\widehat{\boldsymbol{\beta}} - \boldsymbol{\beta})^\top \mid \boldsymbol{y}_i\right] &= \left(\boldsymbol{I}_{\boldsymbol{\beta}\boldsymbol{\beta}}\right)^{-1} + O_p\left(m^{-3/2}\right), \\ \mathrm{E}\left[(\widehat{\boldsymbol{\theta}} - \boldsymbol{\theta})(\widehat{\boldsymbol{\theta}} - \boldsymbol{\theta})^\top \mid \boldsymbol{y}_i\right] &= \left(\boldsymbol{I}_{\boldsymbol{\theta}\boldsymbol{\theta}}\right)^{-1} + O_p\left(m^{-3/2}\right), \\ \mathrm{E}\left[(\widehat{\boldsymbol{\beta}} - \boldsymbol{\beta})(\widehat{\boldsymbol{\theta}} - \boldsymbol{\theta})^\top \mid \boldsymbol{y}_i\right] &= O_p\left(m^{-3/2}\right). \end{aligned}$$

where

$$I_{\beta\beta} = \frac{\tau_1}{\tau_2}\sum_{i=1}^{m}\frac{n_i+\tau_1}{n_i+\tau_1+2}\left\{\sum_{j=1}^{n_i}\left(x_{ij}-\overline{x}_i\right)\left(x_{ij}-\overline{x}_i\right)^\top + n_i\gamma_i\overline{x}_i\overline{x}_i^\top\right\}$$

and

$$I_{\theta\theta} = \begin{pmatrix} I_{\lambda\lambda} & I_{\lambda\tau_1} & I_{\lambda\tau_2} \\ I_{\lambda\tau_1} & I_{\tau_1\tau_1} & I_{\tau_1\tau_2} \\ I_{\lambda\tau_2} & I_{\tau_1\tau_2} & I_{\tau_2\tau_2} \end{pmatrix}$$

with

$$2I_{\lambda\lambda} = \sum_{i=1}^{m}\frac{(n_i+\tau_1-1)\,n_i^2\gamma_i^2}{n_i+\tau_1+2}, \quad 2I_{\lambda\tau_1} = -\sum_{i=1}^{m}\frac{n_i\gamma_i}{n_i+\tau_1},$$

$$2I_{\lambda\tau_2} = \frac{\tau_1}{\tau_2}\sum_{i=1}^{m}\frac{n_i\gamma_i}{n_i+\tau_1+2}, \quad 2I_{\tau_1\tau_1} = \frac{1}{2}\sum_{i=1}^{m}\left\{\psi'\left(\frac{\tau_1}{2}\right)-\psi'\left(\frac{n_i+\tau_1}{2}\right)\right\},$$

$$2I_{\tau_1\tau_2} = -\frac{1}{\tau_2}\sum_{i=1}^{m}\frac{n_i}{n_i+\tau_1}, \quad 2I_{\tau_2\tau_2} = \frac{\tau_1}{\tau_2^2}\sum_{i=1}^{m}\frac{n_i}{n_i+\tau_1+2}.$$

Here, $\psi'(a)$ is the derivative of the digamma function, $\psi(a) = \Gamma'(a)/\Gamma(a)$. Based on the parameter estimates, the empirical best predictor of $\xi_i = c_i^\top\beta + v_i$ for some fixed vector c_i is $\widehat{\xi}_i = c_i^\top\widehat{\beta} + \widetilde{v}_i(\widehat{\beta},\widehat{\lambda})$. Kubokawa et al. (2016) derived the asymptotic approximation of the MSE of $\widehat{\xi}_i$ as follow:

$$\mathrm{E}[(\widehat{\xi}_i-\xi_i)^2] = \frac{1-\gamma_i}{n_i}\frac{\tau_2}{\tau_1-2} + \gamma_i^2 c_i^\top\left(I_{\beta\beta}\right)^{-1}c_i + n_i\gamma_i^3\frac{\tau_2}{\tau_1-2}I^{\lambda\lambda} + O\left(m^{-3/2}\right),$$

where $I^{\lambda\lambda}$ is the (1, 1)-element of $I_{\theta\theta}^{-1}$.

As an alternative modeling strategy for heteroscedastic variance, Sugasawa and Kubokawa (2017b) proposed the model:

$$\mathrm{Var}(v_i) = \tau^2, \quad \mathrm{Var}(\varepsilon_{ij}) = \psi(z_{ij}^\top\gamma),$$

where z_{ij} is a sub-vector of x_{ij}, γ is a vector of unknown parameters and $\psi(\cdot)$ is a known positive-valued function such as $\exp(\cdot)$. A notable feature of the model is that it does not assume normality for the random effect and error terms. Under the settings, β is estimated by the generalized least squares estimator

$$\widetilde{\beta}\left(\tau^2,\gamma\right) = \left(\sum_{i=1}^{m}\mathbf{X}_i^\top\mathbf{\Sigma}_i^{-1}\mathbf{X}_i\right)^{-1}\sum_{i=1}^{m}\mathbf{X}_i^\top\mathbf{\Sigma}_i^{-1}\mathbf{y}_i.$$

and an estimator of τ^2 derived from a moment-based method is

$$\widetilde{\tau}^2(\boldsymbol{\gamma}) = \frac{1}{N}\sum_{i=1}^{m}\sum_{j=1}^{n_i}\left\{\left(y_{ij} - \boldsymbol{x}_{ij}^{\top}\widehat{\boldsymbol{\beta}}_{\mathrm{OLS}}\right)^2 - \sigma^2\left(\boldsymbol{z}_{ij}^{\top}\boldsymbol{\gamma}\right)\right\},$$

where $\widehat{\boldsymbol{\beta}}_{\mathrm{OLS}} = \left(\sum_{i=1}^{m}\boldsymbol{X}_i^{\top}\boldsymbol{X}_i\right)^{-1}\sum_{i=1}^{m}\boldsymbol{X}_i^{\top}\boldsymbol{y}_i$. Furthermore, $\boldsymbol{\gamma}$ is estimated by the following estimating equation:

$$\begin{aligned}\frac{1}{N}\sum_{i=1}^{m}\sum_{j=1}^{n_i}\Bigg[&\left\{y_{ij} - \bar{y}_i - \left(\boldsymbol{x}_{ij} - \overline{\boldsymbol{x}}_i\right)^{\top}\widehat{\boldsymbol{\beta}}_{\mathrm{OLS}}\right\}^2 \\ &- \left(1 - 2n_i^{-1}\right)\sigma^2\left(\boldsymbol{z}_{ij}^{\top}\gamma\right) - n_i^{-2}\sum_{h=1}^{n_i}\sigma^2\left(\boldsymbol{z}_{ih}^{\top}\boldsymbol{\gamma}\right)\Bigg]z_{ij} = 0.\end{aligned} \tag{8.12}$$

Note that the estimating Eq. (8.12) does not depend on $\boldsymbol{\beta}$ and τ^2. Hence, the parameter estimation consists of three steps; first, solve the estimating Eq. (8.12) to get $\widehat{\boldsymbol{\gamma}}$, then obtain $\widehat{\tau}^2 = \widetilde{\tau}^2(\widehat{\boldsymbol{\gamma}})$ and $\widehat{\boldsymbol{\beta}} = \widetilde{\boldsymbol{\beta}}(\widehat{\tau}^2, \widehat{\boldsymbol{\gamma}})$. The EBLUP of $\xi_i = \boldsymbol{c}_i^{\top}\boldsymbol{\beta} + v_i$ for some fixed vector $\boldsymbol{c}_i$ is

$$\widehat{\xi}_i = \boldsymbol{c}_i^{\top}\widehat{\boldsymbol{\beta}} + \sum_{j=1}^{n_i}\frac{\widehat{\tau}^2\widehat{\sigma}_{ij}^{-2}}{1 + \widehat{\tau}^2\sum_{\ell=1}^{n_i}\widehat{\sigma}_{i\ell}^{-2}}(y_{ij} - \boldsymbol{x}_{ij}^{\top}\widehat{\boldsymbol{\beta}}),$$

where $\widehat{\sigma}_{ij}^2 = \psi(\boldsymbol{z}_{ij}^{\top}\widehat{\boldsymbol{\gamma}})$. Under some regularity conditions (e.g., existence of finite higher moments of v_i and ε_{ij}), Sugasawa and Kubokawa (2017a) established asymptotic properties of the estimator and derived asymptotic approximation of the MSE of the EBLUP $\widehat{\xi}_i$.

References

Arima S, Datta GS, Liseo B (2015) Bayesian estimators for small area models when auxiliary information is measured with error. Scand J Stat 42:518–529

Carvalho CM, Polson NG, Scott JG (2010) The horseshoe estimator for sparse signals. Biometrika 97:465–480

Chakraborty A, Datta GS, Mandal A (2016) A two-component normal mixture alternative to the Fay-Herriot model. Stat Transit New Ser 17:67–90

Dass SC, Maiti T, Ren H, Sinha S (2012) Confidence interval estimation of small area parameters shrinking both means and variances. Surv Meth 38:173–187

Datta J, Dunson DV (2016) Bayesian inference on quasi-sparse count data. Biometrika 103:971–983

Datta GS, Hall P, Mandal A (2011) Model selection by testing for the presence of small-area effects, and application to area-level data. J Am Stat Assoc 106:362–374

Datta GS, Mandal A (2015) Small area estimation with uncertain random effects. J Am Stat Assoc 110:1735–1744

Ghosh M, Sinha K, Kim D (2006) Empirical and hierarchical Bayesian estimation in finite population sampling under structural measurement error models. Scand J Stat 33:591–608

Hamura H, Irie K, Sugasawa S (2022) On global-local shrinkage priors for count data. Bayesian Anal 17:545–564

Jiang J, Nguyen T (2012) Small area estimation via heteroscedastic nested-error regression. Can J Stat 40:588–603

Kubokawa T, Sugasawa S, Ghosh M, Chaudhuri S (2016) Prediction in heteroscedastic nested error regression models with random dispersions. Stat Sin 26:465–492

Maiti T, Ren H, Sinha A (2014) Prediction error of small area predictors shrinking both means and variances. Scand J Stat 41:775–790

Molina I, Rao JNK, Datta GS (2015) Small area estimation under a Fay-Herriot model with preliminary testing for the presence of random area effects. Surv Meth 41:1–19

Opsomer JD, Claeskens G, Ranalli MG, Kauermann G, Breidt FJ (2008) Non-parametric small area estimation using penalized spline regression. J Roy Stat Soc B 70:265–286

Scott JG, Berger JO (2006) An exploration of aspects of Bayesian multiple testing. J Stat Plan Inference 136:2144–2162

Sugasawa S, Kubokawa T (2017a) Bayesian estimators in uncertain nested error regression models. J Multivar Anal 153:52–63

Sugasawa S, Kubokawa T (2017b) Heteroscedastic nested error regression models with variance functions. Stat Sin 27:1101–1123

Sugasawa S, Kubokawa T, Ogasawara K (2017) Empirical uncertain Bayes methods in area-level models. Scand J Stat 44:684–706

Sugasawa S, Tamae H, Kubokawa T (2017) Bayesian estimators for small area models shrinking both means and variances. Scand J Stat 44:150–167

Tang X, Ghosh M, Ha NS, Sedransk J (2018) Modeling random effects using global-local shrinkage priors in small area estimation. J Am Stat Assoc 113:1476–1489

Torabi M, Datta GS, Rao JNK (2009) Empirical Bayes estimation of small area means under a nested error linear regression model with measurement errors in the covariates. Scand J Stat 36:355–368

Wang J, Fuller W (2003) The mean squared error of small area predictors constructed with estimated error variances. J Am Stat Assoc 98:716–723

Ybarra LMR, Lohr SL (2008) Small area estimation when auxiliary information is measured with error. Biometrika 95:919–931

You Y, Chapman B (2006) Small area estimation using area level models and estimated sampling variances. Surv Meth 32:97–103

Printed in the United States
by Baker & Taylor Publisher Services